# THE CASE OF THE
# FEMALE ORGASM

# THE CASE
# OF THE FEMALE ORGASM

## Bias in the Science of Evolution

Elisabeth A. Lloyd

Harvard University Press
Cambridge, Massachusetts
London, England
2005

Copyright © 2005 by the President and
Fellows of Harvard College
All rights reserved
Printed in the United States of America

*Library of Congress Cataloging-in-Publication Data*

Lloyd, Elisabeth Anne.
The case of the female orgasm : bias in the science of evolution /
Elisabeth A. Lloyd.
p. cm.
Includes bibliographical references and index.
ISBN 0-674-01706-4 (cloth : alk. paper)
1. Female orgasm.  2. Human evolution.  I. Title.
QP251.L56  2005
613.9'6'082—dc22     2004059695

*To the memory of Stephen Jay Gould,*
*and to my parents,*
*Stuart Phinney Lloyd and Ruth Sorenson Lloyd*

# Contents

# CHAPTER 1

# Introduction

This is a book about the evolution of female orgasm. Female orgasm is a source of fascination for groups ranging from sex researchers to the lay public, and evolutionists are no exception. A number of evolutionary accounts have been given for the trait of human female orgasm, and I review all 21 available accounts in this book. I shall argue that most of the available evolutionary explanations of female orgasm suffer from one or more debilitating faults, either in the way that they treat evidence or in the way that they are constructed. The path to this conclusion lies in examining the supporting and contravening evidence for the various evolutionary explanations that have been proffered.

This is also a book about biases in evolutionary explanations. At least two main biases have negatively affected the evolutionary explanations of female orgasm. One of them is the bias of assuming that the female orgasm evolved to its present form in human beings because it contributed to the reproductive success of its possessor in some way. It may seem to be obvious that the female orgasm has an evolutionary function, but the obviousness of this conclusion must be reevaluated after looking at the relevant evidence. It turns out that no one has ever adequately shown a function for the female orgasm in increasing either fertility or reproductive success. The second bias is that of being male-centered (androcentric) or of assuming that female sexuality is like male sexuality. There have been various trends over the years away from or toward the view that men and women have similar sexual responses, but, I shall argue, the overwhelming evidence is that women respond differently to the act of intercourse than men do (Lloyd 1993). This fact undermines a number of the evolutionary accounts given for female orgasm, as I will show.

After considering the varied evidence for and against detailed evolutionary accounts of female orgasm, I suggest an analysis, which is intended to make sense of how things could have gone awry with these scientific explanations. I emphasize the bias of adaptationism (assuming that the trait was naturally selected because it has a specific function) and the bias of androcentrism (assuming that female sexual response is like male sexual response to the same stimuli). In my research, I did find an explanation that seems to me to be the best supported by the evidence available today. It assumes neither an adaptive function for female orgasm nor an androcentric account of it. One puzzle facing us today is why evolutionists in general have not accepted this account. And that is the puzzle I address in the final chapter of the book. The analysis there shows that there is a way out of the problems which anyone seeking an explanation of female orgasm encounters. In sum, this is a book in which I attempt to discern the best available evolutionary explanation of female orgasm, and in which I diagnose some of the problems that have faced this investigation in the past.

## Evolutionary Accounts

The challenge in providing an evolutionary explanation is to give a historical account of how a group of organisms came to have a particular property or trait. Adaptive accounts, in which the trait in question is assumed to have resulted from the direct operation of natural selection, are the most prominent sort of evolutionary explanations. On such accounts the traits themselves are called "adaptations," and the explanatory game is to provide a description of the past history of the environment and the organism that reveals how the possession of the trait has contributed to the reproductive success of an organism that has the trait—or to its relatives' reproductive success. (Throughout the book, I shall use "reproductive success" instead of "inclusive fitness.")

Take the giant anteater of South America, *Myrmecophaga tridactyla*. This is an animal belonging to an order of mammals, Edentata, which includes armadillos and sloths as well as anteaters. They all descended from an animal much like an armadillo, minus the armor. Anteaters have since evolved special traits that enhance their survival and reproduction—adaptations. Their special adaptations for a lifestyle of consuming ants include specializations of the head and throat anatomy. The giant anteater has a tongue up to 60 centimeters long, which it can extend an entire head length beyond its body while it is eating the ants from a nest. Along with this long tongue, the anteater has specialized giant salivary glands, to keep the tongue covered with sticky saliva to pick up the ants. Further adaptations include the connection of the tongue to specialized hyoid bones in the neck, which help guide and stretch the long tongue. In addition, the face of the anteater is elongated, with a long snout that is suitable for poking into an ant nest. Each of these traits was specially shaped by natural selection over time. Those animals with the more specialized jaws and longer tongues were more likely than those animals with fewer or less extreme specializations to survive and reproduce. Repeating this process of selection over evolutionary time led to the evolution of the animals from their armadillo-like ancestors to the highly adapted giant anteater that lives today. (Details on the anteaters and the order, Edentata, can be found in Moeller 1975, Naples 1999, and Rose 2001.)

## Adaptations

As Mary Jane West-Eberhard writes, "it is correct to consider a character an 'adaptation' for a particular task only if there is some evidence that it has evolved (been modified during its evolutionary history) in specific ways to make it more effective in the performance of that task, and that the change has occurred due to the increased fitness that results" (West-Eberhard 1992, p. 13; I present a contrasting view of adaptation in Chapter 6). In other words, adaptations are traits that have evolved to serve a particular fitness-enhancing role, and that is why they are prevalent. Demonstrating that a trait is an adaptation, though, is a complicated affair.

There are a number of types of evidence that are relevant for showing that a trait is an adaptation. Barry Sinervo and Alexandra Basolo (1996, p. 150) serves as a useful summary of the state of the art in documenting adaptations. I will use their four-part analysis as a convenient vehicle for discussing the various requirements for adaptations in the literature. Sinervo and Basolo have the virtue of compiling the requirements stated by others for establishing that something is an adaptation. I shall list and describe those types of evidence here.

First, it should be shown that individual or geographic variations in a trait have a genetic basis (Falconer 1981; Sinervo and Basolo 1996, p. 150; Reznick and Travis 1996, p. 261; Griffiths et al. 2002). This is desirable because adaptation by natural selection assumes that different phenotypes, or variants of a trait, will, for the most part, have different genetic underpinnings (but see Pigliucci 2001 and West-Eberhard 2003). This makes it possible for the population to change its rates of the specific variants of the trait over time. A genetic basis is a prerequisite of natural selection.

Second, the trait should be shown to influence reproductive success (Lande and Arnold 1983; Endler 1986). Without a connection between reproductive success and the trait, there is no basis for con-cluding that the trait evolved over evolutionary time by natural selection. There are a number of methods available for showing that a particular trait is correlated with higher or lower reproductive success than its variants. This is *the* crucial element in showing that selection may have acted on a trait, and will prove to be an important touchstone in my examinations of the adaptive significance of female orgasm. One of the most popular ways is to perform studies of the design and efficiency of a trait, in order to see whether having the specific variant of the trait enhances the performance of the organism in some way (Larson and Losos 1996, p. 201; Lauder 1996; Seger and Stubblefield 1996). In this approach, the researchers measure the performance of a trait in undertaking a specific task, say, a beak cracking seeds of a certain size, and compare that performance with organisms having a different variant of the trait. They also compare the current manifestation of the trait—especially anatomical traits—with ancestral manifestations of the trait, thus showing an evolutionary change, presumably in response to selection pressures.

An additional aspect of this type of evidence is indication that the ancestral environment really did exert the selection pressure. Evidence for ancestral selection pressure can be obtained in a number of ways. One way is to extrapolate back from current selection pressure, although this is risky, because of evolutionary changes in the biotic and abiotic environment. Another method is to locate paleobotanical or climatic geological information about past environments. This evidence has the virtue of being independent of specific hypotheses about adaptation, and can provide strong support for an ancestral selection scenario.

Another important method for showing that variation in a trait has an impact on reproductive success is the "mark-recapture" method used in ecological genetics (Endler 1986; Reznick and Travis 1996). By capturing individual organisms in the wild, measuring their trait value, releasing them, and later recapturing them, researchers can estimate mortality rates and lifetime reproductive success. Historically, such findings have proved an important source of links between trait values and reproductive success.

A third requirement of evidence for an adaptation is that a mechanistic account explaining the links between the trait and reproductive success in the wild should be elucidated (Ferguson and Fox 1984; Grant 1986; Jayne and Bennett 1990; Arnold 1983; Levins and Lewontin 1985; Reznick and Travis 1996). In other words, we need to know the mechanism by which having a particular variant of a trait actually manages to contribute to the differential reproductive success of its owner. An excellent example of this kind of evidence can be found in the human sickle cell case, to be reviewed in a moment. In that case, it was shown that sickled red blood cells were actually resistant to infection by the malarial parasite. Or in the case of the Galápagos finches' beaks, certain beak shapes were shown mechanically to be better at forcing different sizes of seeds open (Grant 1986). Often, this sort of research involves detailed morphological study of the variants of the trait, combined with an engineering analysis of which designs would be most effective and efficient (Seger and Stubblefield 1996). Such studies attempt to establish two things: the correlation of certain values of the trait with reproductive success and the analysis of the mechanisms underlying such success. Problems with this type of approach are analyzed by George Lauder (1996).

The fourth type of evidence for adaptation concerns experiments. The hypothesis that variation in the trait is adaptive should be independently confirmed by experimentally manipulating the selective environment or the trait itself (Endler 1986; Mitchell-Olds and Shaw 1987; Schluter 1988; Wade and Kalisz 1990; Reznick and Travis 1996). Such studies attempt to establish that in different environments, different variants of a trait are more successful reproductively than others, or that in a single environment, different variants of a trait have different reproductive success. This helps nail down the connection between a specific trait and reproductive success. David Reznick and Joseph Travis (1996) elaborate on a couple of examples of this type of study in fish. John Endler (1986) presents a host of additional examples. Note that this kind of study may not be performed on human beings, who cannot ethically be subjected to the kind of manipulations involved. Thus this criterion is not available to those studying the evolution of female orgasm. The closest researchers can come to such tests is simply to analyze "natural experiments," in which variants in the trait are present in the population, and to try to connect these variants to differences in reproductive success and environment. Fundamentally, this amounts to fulfilling the second criterion. It is related to doing comparative studies of populations in different areas and under different conditions. This can help show that populations undergoing hypothesized different selection pressures have different variants of the trait.

In sum, obtaining evidence that a trait is an evolutionary adaptation—that it evolved to serve a particular fitness-enhancing function—requires diligent and creative investigation. Numerous types of evidence are relevant, and the more types of evidence are adduced in support of some particular trait's being an adaptation, the more we can be confident that it really was selected over evolutionary time for performing a fitness-enhancing function.

## Human Examples

Adaptive approaches have proven very fruitful in examining human traits. Here I offer two well-confirmed cases. Genes of the major histocompatibility (MHC) locus are very diverse, and can influence a variety of immune functions, including resistance to infectious and autoimmune diseases, as well as influencing personal odors and pregnancy outcome. Of interest for my discussion is that individuals with a mixed MHC genotype, coding for a variety of proteins, have an advantage over homozygotes against infectious disease. The fact that individuals with different MHC genes have different odors is also relevant to my argument.

It seems that female animals either find more attractive or prefer to mate with males whose MHC genes are dissimilar to those of the females in question. Females also prefer heterozygous males, as shown by studies of mice (Egid and Brown 1989; Boyse et al. 1990; Potts et al. 1991; Penn and Potts 1999), salmon (Landry et al. 2001), and sticklebacks (Reusch et al. 2001). In a seminal paper, Claus Wedekind and colleagues (1995) presented evidence that human females, too, may select their mates on the basis of MHC loci. The mechanism by which this choice occurred involved dissimilarity of smell that is linked with the MHC locus. Forty-nine female students were exposed to T-shirts worn by various men (deodorants and perfumed laundry detergents were eliminated). Both the males and the females were typed for MHC loci. Women were exposed to seven T-shirts each, three of them worn by men with differing MHC loci, three by identical loci, and one control, in a blind study. Women scored the T-shirts for intensity, for pleasantness, and for sexiness. The women were tested when they were believed to be ovulating, since women appear to be most odor sensitive at that time. Each male odor was scored by two or more females, one of similar and one of dissimilar MHC type.

Among females, the scores for sexiness were highly correlated with scores for pleasantness. It was found that the pleasantness of men's body odors scored by different women depends on their respective MHC. In particular, males with *different* MHC loci were perceived as more pleasant than those males with similar MHC loci. Moreover, odors of MHC-dissimilar men reminded the females of their own mates or ex-mates twice as often as those of MHC-similar men. "This indicates that MHC-dependent body odor preferences play a role in actual mate choice," write Wedekind and colleagues (1995).

In another set of evidence, "Couples who had not achieved a rec-ognizable pregnancy after two or more attempts of in vitro fertil-ization (IVF) or tubal embryo transfer (TET) shared a signifi-cantly greater number of [MHC loci] than did control couples who achieved a viable pregnancy with their first IVF or TET cycle" (Wedekind et al. 1995, p. 247; Freeman and Herron 1998). In addi-tion, couples who suffer from recurrent spontaneous abortions also share a higher proportion of the MHC than control groups in many different populations (Beer et al. 1985; Bolis et al. 1985; Thomas et al. 1985; Karl et al. 1989; Ho et al. 1990; Koyama et al. 1991; Ober 1992; Laitinen 1993; Eberhard and Cordero 1995; Jennions 1997; Birkhead 1998).

The hypothesis is made still more plausible by many examples establishing strong links between the nose, physiology, and emo-tions (Stoddart 1991; Potts 2002). There are nevertheless appar-ent problems with this result. Several studies found that females preferred MHC-similar odors, in contrast to the Wedekind study (Beauchamp et al. 1988; Ecklund et al. 1991; Ecklund 1997; Jacob et al. 2002). These studies, however, have been critiqued on statisti-cal grounds (Penn and Potts 1999). In addition and more specifi-cally, women taking oral contraceptives prefer MHC-similar men in experiments challenging the experiments in which females were found to prefer MHC-dissimilar males, as do female mice choosing females as communal nesting and nursing partners. But the latter may have to do with kin selection in raising offspring. In situations where interacting with kin, such as communal nesting behavior, comes into play, selection of similar MHC loci are preferred (Potts 2002, p. 131).

One final piece of evidence comes from a "natural" experiment. Carole Ober and colleagues (1997) also found evidence of MHC-dependent mating populations in a group of Hutterites. The Hut-terites migrated from Europe to North America in the 1870s with a population of about 400 members. Mating occurred almost exclu-

sively within this group. They have maintained scrupulous genea-
logical records since they moved to North America, and thus we
have a very good record of genetic inheritance. They are a highly in-
bred group of human beings, with a relative lack of genetic diversity
in MHC. Nevertheless, there is a relatively lower rate of MHC
homozygotes at birth, which implies selective mating for different
MHC loci on the part of the females, or an increased abortion rate
of MHC-similar fetuses. In addition, a study of Hutterite married
couples showed that the couples were less likely to share MHC
genes than chance would predict, which also indicates preferential
mating of females to males with different MHC genes.

How does this evidence fit with the four standards for establishing adaptations? Note that in the MHC study, three of the desider-
ata for showing that a trait is an adaptation are at least preliminar-
ily met. First, there is a documented link between a phenotypic
trait, smell, a physiological trait, MHC locus, and its genetic underpinnings. Second, there seems to be evidence linking the different
variants of the trait with reproductive success, especially with regard to spontaneous abortion. Third, the case seems to meet the requirement that there is a known mechanism linking the trait and
reproductive success. Thus the case meets three of the four criteria
for showing that a trait is an adaptation. Overall, therefore, we
have good reason to regard the choice of males based on smell associated with the MHC loci as an adaptation of human behavior. This
example is especially important because it involves a social trait—
mate choice. I consider the female orgasm to be a physiological
trait, and not a social trait, although this is debatable. Most of the
examples of human adaptations are physiological, including the
best example, which follows.

In another example of adaptation, human beings show a relatively high percentage of a variant of the hemoglobin molecule,
called the S (sickling) hemoglobin (Cavalli-Sforza and Bodmer
1971; Templeton 1982). The reason that this is a puzzle is that any

individual who is homozygous for S hemoglobin has a variety of serious health problems, including lowered fertility and high preadult mortality (Templeton 1982, p. 16). So why is the S allele at a substantial frequency in the human population? The answer lies in the high fitness value of the mixed hemoglobin, one allele for A (normal) hemoglobin and one for S hemoglobin. Such a combination provides protection against malaria, a widespread killer.[1]

There is a close geographic association between the distribution of malaria and the sickle trait, which supports the concept that the sickle cell trait may have a survival advantage against some adverse condition in the tropics (Mackey and Vivarelli 1952; Allison 1954; Serjeant 2001). It was Anthony Allison who proposed that persons with one sickle cell gene developed malaria less frequently and less severely than those without the trait. It is now accepted that the sickle cell trait confers some protection against malaria during a critical period of early childhood (Allison 1957; Rucknagel and Neel 1961; Motulsky 1964; Templeton, 1982, p. 16, comments that most malarial deaths occur in infants and children).

Distribution of the sickle cell mutation and survival advantage during a malarial infestation determines the geographic distribution of the sickle cell gene. It seems to have evolved four times (Serjeant 2001), all in tropical areas prone to malarial infestation. There is also an Asian version, predominant in Saudi Arabia and central India, areas that are also subject to malarial infestation.

A relation between the genetics of hemoglobin and the sickle cell trait was established by Linus Pauling and colleagues (1949). The molecular change is an amino acid substitution. Thus the link between a genetic basis and the phenotypic trait is established.

It is also important that carriers (heterozygotes) of the sickle trait have relatively normal hemoglobin function, while homozygotes of the sickle cell gene develop a severe form of anemia, which is accompanied by relatively low fitness.

The sickle cell trait is a prime example of a human adaptation to challenging circumstances. The cost of the SS genotype is out-weighed by the advantage of the SA, or sickle cell carrier, in malarial areas, which are widespread. The sickling gene is very rare in parts of the world where malaria does not appear (Templeton 1982, pp. 18–19). The sickle cell case is also the best, most elaborated case of a human adaptation. The case for the genetic determination of the phenotypic trait is very well established; and that fulfills the first requirement outlined by Sinervo and Basolo (1996). In addition, the case for the correlation between the mixed genotype and malarial resistance is also well established; and that ties the trait into reproductive success. Moreover, the sickling gene is found in populations in proportion to the extent to which the group is exposed to malarial infection; and that establishes the environmental connection to the genotype. In fulfilling the third requirement of Sinervo and Basolo, the mechanism of the adaptation has been well elucidated. Thus the sickling case proves to be a model of adaptationist thinking, linking as it does genes with reproductive success in an environment.

## Alternative Historical Accounts

Other types of historical account besides adaptationist ones are also available to evolutionists; these focus on the interrelation of an organism's multiple traits and on the development of the organism. The attraction of adaptive explanations lies in their appeal to the shaping power of natural selection, which is the dominant force of evolutionary change. In addition, adaptive explanations are often subject to testing. But Charles Darwin himself recognized the limits of adaptive explanations, as is clear from his emphasis on what he called the "correlation of characters"; he argued that some traits must be explained through their relation to other traits that might be directly selected. It is also possible that traits may not have arisen for the role that they currently serve in selection, but may have arisen due to the way that animals grow.

Consider Darwin's example of such a trait. "The sutures in the skulls of young mammals have been advanced as a beautiful adaptation for aiding parturition, and no doubt they facilitate, or may be indispensable for this act; but as sutures occur in the skulls of young birds and reptiles, which have only to escape from a broken egg, we may infer that this structure has arisen from the laws of growth, and has been taken advantage of in the parturition of the higher animals" (Darwin 1964, p. 197). Here is a trait that is important today for animals that give birth to live offspring, because the skull sutures allow the skull to deform so that it can pass through the narrow birth canal. Nevertheless, sutures are not an *adaptation* for that job, because they appeared in the lineage before that use was ever made of them.

Or consider the interesting and familiar trait of the male nipple in mammals. Why do male mammals have nipples? Clearly, nipples are necessary to the reproductive success of any female mammal (including human beings, in evolutionary time). Therefore there would be strong selection pressures on female nipples—to ensure that they are functional feeding devices, connected to the milk-producing breast tissue in the right way, and even to ensure that nursing is sometimes pleasurable by connecting the nipple to pleasurable and even sexually stimulating sensations to reward the female for nursing. Before the male embryo splits off from the primordial (female) basic type of human embryo, directions for building nipples are already activated. Thus female and male nipples are technically "homologues." This is granted even by those who have recently contested the notion of homology in development. Thus males get nipples as a developmental or embryological consequence of females needing to have them. And the trait persists among males because it is consistently selected in females. But, strictly speaking, nipples are not adaptations in men; they are byproducts of adaptive pressures on women.

Such "nonadaptive" explanations are now seen by many biologists as inferior, even though the explanations may provide detailed scientific accounts of a trait's existence. Such biologists, called "adaptationists" because of their commitment to finding adaptive explanations of a trait, are sometimes hesitant to allow that other accounts may be superior in particular cases. This issue comes into play in the case of female orgasm because one of the best-supported accounts is, in fact, an account based on adaptation in the male but not the female.

### Accounts Discussed in This Book

I review the sex research on female orgasm in Chapter 2, focusing on a range of studies that show that sexual intercourse is only sometimes accompanied by female orgasm. Although there are problems with the methodology used in sex research, any evolutionary account must be compatible with such findings, because they are the only scientific results available. In Chapters 3 and 4, I introduce a number of adaptationist accounts of the evolution of female orgasm, accounts that assume that female orgasm is like the sickle cell trait, a special adaptation contributing to the reproductive success of its owner. I find these accounts deficient for several reasons. First, they seem to take little notice of the basic sex research that has been done on female orgasm, mischaracterizing the nature of the very trait whose existence they seek to explain. Second, a few of the explanations display flawed reasoning: if we think about the implications of the account, the scenario it offers simply would not result in evolution of the trait. Finally, many of them equate female orgasm with intercourse. But this can only be a confusion, given the basic findings reported in Chapter 2, that intercourse is often not accompanied by orgasm—even when the woman herself is capable of orgasm—and that orgasm is often not accompanied by intercourse.

Thus I find all of the explanations discussed in Chapters 3 and 4 defective for not taking account of the sex research, not making logical sense, or reducing orgasm merely to orgasm with intercourse. It is important to note that I have not cherry-picked the worst explanations for display in these chapters, for my approach in the book has been to address every single explanation of the evolution of female orgasm that I could find. Thus my analysis includes work by the most prominent researchers on this topic, people such as Frank Beach, the pioneer of the effect of sex hormones on behavior; Sarah Blaffer Hrdy, one of today's premier primatologists; John Alcock and Richard Alexander, leading sociobiologists; and Bernard Campbell, who was a leading expert on human evolution. I should note that the problems I found with these explanations of female orgasm are not necessarily shared by accounts of other traits in human evolution. My study is meant only to establish that there are problems with the evolutionary accounts of the human trait considered here, female orgasm, and not to impugn the field of human evolution as a whole.

Readers may also wonder if, in these discussions in Chapters 3 and 4, I am applying standards of evidence or explanation that are too harsh. Perhaps I am holding the evolutionists to too high a standard, one that is not required in explanations of other traits. This, however, is not the case. In each chapter I discuss the standards of evidence that I use, and defend my position that these are shared standards of evidence that are applicable to all evolutionary explanations.

In Chapter 5, I address what I judge to be the best available explanation for the evolution of female orgasm: that proposed by Donald Symons. I do not take the theory to be beyond dispute, but it is in accord with the evidence we have today and has the quality of unifying some otherwise mysterious findings. The explanation is not an adaptationist explanation at all; it is based on a notion of shared development between male and female sex organs, in parallel to the case of the developmental origin of male nipples.

The Symons account is not without its critics, and a number of objections have been brought against it, which I discuss in Chapters 5 and 6. One of the main confusions about this view revolves around the notion of *adaptation* itself; when is it appropriate to say that something is or is not an adaptation? There has been a heated debate in the evolutionary literature concerning this question. A proper treatment of the issues would take another entire book, but I discuss these issues at length in Chapter 6. Amidst the contention about adaptationism I find a near consensus that a trait, in order to be an adaptation, must be linked to reproductive success: there must be a connection between having the trait and doing better reproductively than those without the trait. But that has not been shown for female orgasm: women can and do get pregnant easily without having orgasm during copulation, and there is no evidence that shows that orgasm makes a difference to fertility or reproductive success. This is the topic I take up in Chapter 7, when I consider the recent explanations of female orgasm based on the concepts of sperm competition.

Sperm-competition accounts have become remarkably popular, and are cited in lay and scientific works alike (Mah and Binik 2001, p. 827; Zuk 2002, p. 147; Andrews et al. 2002, p. 17; Fausto-Sterling et al. 1997, pp. 404, 416). Some cite these accounts completely uncritically: Buss 1994, pp. 75–76; Singh et al. 1998; Daly and Wilson 1999; Jolly 2001, pp. 90–91; Rodgers 2001, pp. 323–329; Barrett et al. 2002, pp. 114–115; Singh et al. 2002. Such accounts are based on the idea that the uterus develops suction after orgasm, and that this results in semen and sperm being sucked into the uterus. Research on this idea dates back to the 1950s, and the issue is still a matter of contention. I review the controversy in the literature, and then consider a recent study on whether sperm is sucked into the cervix and uterus. This study also offers an adaptive account of female orgasm, based on the concept that orgasm helps the female control whether the sperm from a particular male will fertilize her egg or not. Unfortunately, this study presents such a confused and faulty set of statistical supports that it must be considered unreliable. Once these statistical problems are made clear, it seems amazing that so many people have accepted the study, and that this state of affairs has continued for over a decade. The end result, though, is that there is still no evidence for a link between female orgasm and fertility. There is a further account that is based on the statistically flawed account, and it, too, makes unacceptable use of statistics. Its results have also failed to be reproduced in an independent experiment using the same methodology. My conclusion, then, is that there is at present no known connection between female orgasm and reproductive success, although I point to one promising avenue of research that might yet establish a connection along these lines.

Though at this time I find no credible evidence that female orgasm is an adaptation, I am open to such a finding. Female orgasm may very well turn out to be an adaptation, exquisitely designed for some special but obscure function. None of my arguments is meant to rule this possibility out, and I primarily wish to draw attention to the controversy on this topic and the fascinating elements involved in that controversy. I have no predictions about the future regarding whether or not female orgasm will turn out to be an adaptation or not, but wish to emphasize the state of evidence at the present time.

## Analysis and Significance

Chapters 3–7 thus cover the variety of explanations given for the evolution of female orgasm. In the course of the chapters, I raise numerous objections—methodological, logical, and evidential—to these explanations. The basis of these objections lies in an appeal to general, accepted standards of evidence. The question then arises: What happened with most of these explanations? Many of their authors are perfectly familiar with these standards; why did they fail to apply them? This is a question I propose to answer in Chapter 8. There, I argue that certain background assumptions have had a detrimental effect on much of the science thus far produced on the question of female orgasm. More specifically, I argue that a male-centered bias has played a major role in generating some of these evidential problems, while an adaptationist bias bears responsibility for an overlapping set of difficulties. In that chapter I return to the question of whether some evidential standards I use are appropriate, and I argue that they are. I also detail the connections between particular failures of evidential standards and the particular biases that, on my analysis, caused those failures.

Overall, therefore, the book should be seen as a case study on how biases and background assumptions can affect the practice of science. But one may ask, "So what?" What difference does it make that this particular topic seems to attract deficient scientific explanations?

One answer is scientific. I argue on behalf of an account that seems to me to have good evidence supporting it, but that has been rejected on inadequate grounds. The Symons account may not necessarily be correct, but at this time it is, I shall argue, the best-supported account. I find that it has been underappreciated by the researchers interested in the evolution of female orgasm, and set out to correct this. There is also the problem that many people believe the statistically unacceptable studies in which female orgasm is seen as an adaptation, because it is hypothesized to induce a sucking motion of the uterus. I set out to correct this, as well.

Another answer is primarily social. Among the mosaic of behaviors whose social role remains unsettled, few are more debated—more politicized, more contentious—than female sexuality. It has been a tradition, since the inception of evolutionary work on female orgasm in 1966, to incorporate the evolution of female sexuality into general evolutionary accounts of humankind. These general evolutionary accounts help define what it is to be human. They help show how we are unique, or how we differ from our ape ancestors. As has been recognized since the first "man the hunter" evolutionary accounts, the narratives involving our deep history are often used to ground understandings of ourselves today. And this is also true for evolutionary accounts of female orgasm. The implication usually drawn from evolutionary accounts is that they give the "natural" function of the trait in question, and, regardless of its invalidity, many people do make the inference from what is seen as "natural" to what is seen as socially acceptable. Is female orgasm naturally evolved to help us maintain our pair bonds? Then that is its natural function, on the common view, and if a woman's orgasm does not accord well with that natural function then there must be something wrong with the woman—she is being unnatural. Is female orgasm just a byproduct of selection on the male orgasm? Then, on the usual approach, little is implied about the trait's natural function; in fact, it doesn't seem to have one. It really does make a difference to current understandings of female sexuality which of the available evolutionary accounts is taken to be true. On one set of accounts, those many women who do not experience orgasm regularly from intercourse are seen as unnatural or, in Masters and Johnson's terms, "dysfunctional." On other accounts, these same women are seen as functioning normally, and there is no social or psychological judgment against them. Thus the issue of what difference the evolutionary account makes is simply answered: it makes a difference because it affects how women's sexuality is socially and personally perceived and categorized today. I am not endorsing the argument from an evolutionary view about a trait to a normative or moral one, but given that this move is made almost universally outside philosophical circles, we cannot deny its social impact.

A third answer to the "So what?" question involves our understanding of how science is done. There has been much discussion recently, in the philosophical literature and beyond it, about whether or not a science's being male centered makes a difference to the validity of its results. Many have disputed the feminist claim that male-centered science is often deficient science. Throughout the book, I discuss striking examples of hypotheses not matching up with available evidence. I claim that this occurs because of the impact of various background assumptions, especially androcentrism and adaptationism. The presentation of this book as a case study of science gone astray is not meant to imply that all science operates this way. But some have denied that such biases *ever* interfere with the outcome of the science itself, and this book shows that such a view is mistaken.

# CHAPTER 2

# The Basics of Female Orgasm

Although there have been numerous definitions of female orgasm offered, most agree that it involves a sensory-motor reflex including clonic contractions (spasms) of the pelvic and genital muscle groups (see reviews in Levin 1981; Mah and Binik 2001). For example, Alfred Kinsey defined it as "the explosive discharge of neuromuscular tension and the peak of sexual response," while Douglas Mould says orgasm is the "clonic contractions of pelvic and abdominal muscles initiated by a spinal reflex" (cited in Whipple, Hartman, and Fithian 1994, p. 430). My personal favorite, quoted by John Bancroft, the recent director of the Kinsey Institute, describes orgasm as the "combination of waves of a very pleasurable sensation and mounting of tensions, culminating in a fantastic sensation and release of tension" (from Vance and Wagner 1976, cited in Bancroft 1989, p. 81). To make the technical concept more accessible to the general reader, Kline-Graber and Graber (1975) appeal to a parallel with the reflex response that occurs when a doctor taps the knee and the lower leg subsequently kicks. The sensory part of the reflex loop involves the incoming sensory stimulation from the knee. The motor part involves signals outgoing from the spinal cord inducing the muscular contractions that make the leg kick. In the case of female orgasm, the sensory part of the reflex includes input from various muscles and tissues in the genitopelvic area, while the motor

21

part involves the reflex contractions of muscle groups in the genito-pelvic region.

There is a certain amount of disagreement among sex researchers about female orgasm. On my analysis, these disagreements concern exactly which muscles and tissues are involved in each of the sensory and motor arms of the orgasmic reflex. Most studies have emphasized the overriding importance of clitoral stimulation to the occurrence of orgasm, but additional locations of sensory input to the reflex have been suggested. There is evidence that, for some women, stimulation of part of the anterior wall of the vagina, around the twelve o'clock position (sometimes known as the "G-spot"), provides an important alternative or additional source of sensory input for orgasm.[1] Others have suggested that stimulation of the cervix and of the deep tissues surrounding the vagina and uterus are important sources of stimulation, although there has been less experimental support for this claim.[2]

As for the motor part of the orgasmic reflex, there are five pelvic muscle groups involved, and there is also disagreement about which ones are most essential to orgasm. The first muscle group includes the superficial muscles surrounding the vaginal opening. These are particularly important for inducing the drainage of the erectile tissues that become engorged with blood during sexual excitement. The second group of muscles supports the perineum, the tissues surrounding and holding together the pelvic organs. The third set of muscles includes the pubococcygeus muscle, which goes from the pubic bone to the tailbone, and includes the important shelf of muscles surrounding the lower end of the vagina. These muscles can be contracted voluntarily, and are sometime used to enhance sexual stimulation (Kline-Graber and Graber 1975). They underlie the tissues that make up what Masters and Johnson call the "orgasmic platform," which consists mainly of tissues that are engorged with blood (vasocongested) during sexual excitement, and which are drained of blood through the spasms of the muscle groups.[3] The uterus itself makes up the fourth muscular location of the spasms of orgasm, while the muscles surrounding the vagina along its length are the final set of muscles involved in the motor part of the orgasmic reflex.

As we can see, female orgasm turns out to be quite a bit more neurologically complicated than the simple knee-kick reflex. In addition, other systems of the body may also be involved with orgasm. For example, central brain activity, hormones, and neurotransmitters have all been implicated experimentally in the orgasmic response (Mah and Binik 2001 offer a useful summary of the state of research; many of the claims have not been substantiated). In addition, there is an entire set of psychological dimensions to sort out. Several efforts have been made to integrate the biological and psychological perspectives on female orgasm, but none of them is widely accepted.[4] I shall focus on the physiology of the pelvic and genital area only, with some appeal to the neurohormone oxytocin.

For my purposes, in looking at evolutionary explanations, I shall stick to the relatively reductionistic biological descriptions of female orgasm. My focus shall be on the evolution of the reflex loop described and its physiological aspects. Admittedly, this approach can be faulted for perhaps excluding some crucial psychological aspect of orgasm, but it is the most useful for cross-comparison of evolutionary explanations, many of which focus on the orgasm under its narrow biological description, especially in nonhuman primates. In addition, a focus on orgasm itself, to the exclusion of other aspects of sexuality, is also narrow. Nevertheless, orgasm is a significant and crudely quantifiable aspect of sexual response that has caught the attention of evolutionists.

Concerning the pattern of the female orgasm's occurrence, Masters and Johnson's approach is still the most widely used. According to Masters and Johnson, there are four successive phases into which one can divide the process of orgasm. The first, the "excitement" phase, involves the induction of sexual tension or arousal through psychological or physical stimulation. The second phase, the "plateau" phase, represents a heightened level of sexual tension and excitement, from which orgasm can be reached.[5] The third, the "orgasmic" phase encompasses several seconds of involuntary climax, where the sexual tension is relieved in explosive waves of intense pleasure. This phase is often accompanied by rigid muscles (myotonia) as well as by the spasms or contractions of the pelvic muscles that make up orgasm. Finally, there is a "resolution" phase, where the vasocongestion that occurred in the excitement and plateau phases subsides, and the woman returns to the pre-excitement state.

There is a complication in applying the "EPOR" model, as Masters and Johnson's account is called, to female orgasm: that many women have more than one—some as many as a dozen or more—orgasms before reaching a resolution phase (Masters and Johnson 1966, pp. 6–7). According to Masters and Johnson, these women return to the plateau phase of excitement following orgasm, and do not progress into the resolution phase until after the last orgasm, and even then, the process may be slow. Contrary to popular belief, the ability to have multiple orgasms is not confined to women; prepubescent boys and some young men can also experience multiple orgasms.[6]

At any rate, the existence of multiple orgasms in women, which Kinsey documented at a rate of about 14% (Kinsey et al. 1953, p. 375), complicates the physiological profile of the stages of sexual excitement and fulfillment for women.[7] I shall touch on this issue in my examination of the evolutionary accounts of female orgasm.

## Orgasm with Masturbation

How do women have orgasms? Many of women's orgasms are achieved through masturbation, as both Gebhard and Kinsey and colleagues indicate. The most striking things about female masturbation are how likely it is to produce orgasm and how little it resembles, mechanically, the stimulation received from intercourse. On the reliability of masturbation to produce orgasm, Kinsey and colleagues found that 62% of their sample of 5,940 women had masturbated, and among these, only 4% did not have orgasm with masturbation (1953, p. 142; Gebhard 1970, p. 15). Similarly, Hite found that 82% of her sample of 1,844 women stated that they masturbated, and only 4% did not have orgasm that way, while 95% of these women "could orgasm easily and readily, whenever they wanted" (1976, p. 59).

Of the methods women use to masturbate, Gebhard notes that the "most common masturbation technique is the manual stimulation of the clitoris and the small lips of the vulva" (1970, p. 15). This technique accounts for 84% of all acts of masturbation among the women the Kinsey team surveyed (Gebhard 1970, p. 15). Less than one fifth of women masturbate by inserting an object or fingers into the vagina, and nearly all of those who do so accompany the action with clitoral stimulation (Kinsey et al. 1953; Gebhard 1970, p. 16). As Kinsey himself noted, women almost never masturbate solely in imitation of the act of intercourse, by inserting something into the vagina (Kinsey et al. 1953, p. 163). Hite found that only 1.5% of women masturbate by vaginal insertion alone (1976, p. 411). Moreover, women's preferences for clitoral and labial stimulation are widely known; Kinsey cites 16 sources in European and American literature, dating from 1885 (1953, p. 158). And in contrast to the median amount of time needed for a woman to have an orgasm from intercourse, women take an average of approximately 4 minutes to achieve orgasm with masturbation, the same period as for a man (Kinsey et al. 1953, p. 163).

## Orgasm with Intercourse

There is one point on which all sex researchers agree: the unpredictability and nonequivalence of female orgasm with intercourse. Because this point is crucial to my examination of a number of existing evolutionary explanations, I shall review it in some detail here. I summarize the results of 32 studies that include data on orgasm frequency with intercourse in Table 1. These studies vary a great deal in the sample population, sample size, and in their methodology. Nevertheless, I shall attempt to summarize their main results in a succinct fashion that does some justice both to their variety and to the clear trends that emerge.

Before continuing, I must mention an important hidden component in the studies reviewed below. Most of the studies take the women's reports of orgasm with intercourse at face value, but there is a considerable problem about what, exactly, "orgasm with intercourse" involves. Does a woman who reaches orgasm during intercourse, but only by manual stimulation of the clitoris, count as having "orgasm with intercourse"? Fisher's supplementary studies of the method of orgasm are of particular interest. Fisher found that, on average, 30% of the time women "require direct stimulation [of the clitoris] to give them the final push necessary to reach orgasm" (1973, p. 193). A full 35% of women "said they require such final direct manual stimulation 50 or more percent of the time to attain orgasm," while "only 20 percent of the women said they never require a final push from manual stimulation to reach orgasm" (1973, p. 193). These findings are consistent with Hite's concerning the necessity of direct clitoral stimulation to reach orgasm. She found that 24% of women can regularly reach orgasm without direct clitoral stimulation by hand, while an additional 19% can regularly reach orgasm if direct clitoral stimulation is applied. It turns out that both Kinsey and colleagues (1953) and Gebhard (1966, 1970) self-consciously used figures for orgasm frequency during intercourse that included orgasm reached through direct manual stimulation of the clitoris (according to coauthor Wardell Pomeroy; Hite 1976, p. 233). As Hurlbert and Apt put it, "it may be normal for women to require sustained clitoral stimulation to be orgasmic during coitus" (1995, p. 21).

Thus most studies' results come with an unanswered question: Were orgasms achieved through "assisted" intercourse included in their statistics? It seems likely that they were, given that the questionnaires did not specify how orgasm with intercourse was achieved. The reason this might be a problem for interpreting orgasm frequency in our species is that there are no data on how widespread the practice of assisted intercourse is across cultures. If the practice is not widespread—and none of the scant cross-cultural studies mentions its use—then the frequencies reviewed for orgasm with intercourse will be much higher for Americans and Europeans than they would for most others. Unfortunately, we have no way of resolving this issue without resorting to fresh, newly designed studies. Thus a further caveat must be added to consideration of these results as cross-culturally representative (further cross-cultural evidence will be considered in Chapter 5). Let me proceed to the table of results.

Sixteen of the studies determined what percentage of the women they sampled stated that they "always" had orgasm with intercourse. The results ranged from 6% to 51%. The mean value of the results from all 16 studies was 25.3%, with a median value of 23.5%. I would not claim that the means and medians I calculate later in the chapter have great evidentiary value, but they are one useful way to summarize the trends in the abundant and heterogeneous findings. I strongly recommend that those interested in the details check the original studies and evaluate for themselves what their results are worth. With regard to the 16 studies just summarized, I would emphasize that 2 of them—the only 2 out of all 32 studies I examined—were done with a probabilistic sampling technique, and thus should be given more weight. These are the results of Stanley (1995), which are from sampling by the group called Mass-Observation of 3,450 women in England, and the results of Laumann and colleagues (1994), a study based on random sampling methods of 1,610 Americans. Their results for women who "always" achieved orgasm were 18% and 28.6%, respectively.

TABLE 1 Studies of orgasm rates with intercourse.

| Researchers | Sample size | Results |
|---|---|---|
| Butler (1976) | 195 | 12% always<br>51% more than half the time<br>24% less than half the time<br>8% never |
| Chesser (1956) | 2,000 | 24% always<br>35% frequently<br>26% sometimes<br>10% rarely<br>5% never |
| Christensen and Hertoft (1980) | 32 | Approx. 25% always<br>Almost 20% not usually<br>20% never had orgasm at all |
| Clifford (1978) | 65 | Note: explicitly includes assisted intercourse<br>11% orgasm rate over 90%<br>31% orgasm rate 51–90%<br>58% orgasm rate up to 50% |
| Dickinson and Beam (1931) | 442 | 40% "yes"<br>2% usually<br>3% by clitoris friction<br>15% sometimes<br>10% rarely<br>4% formerly, not now<br>Not experienced with husband: 26% |
| Fisher (1973) | 285 | 39% always or "nearly always"<br>60% "more irregularly or not at all"<br>5–6% never had an orgasm |
| Frank et al. (1978) | 100 | 46% difficulty in reaching orgasm<br>15% no orgasm |

ABLE 1  *(continued)*

| Researchers | Sample size | Results |
|---|---|---|
| Gebhard (1966) | 1,026 | 59% always or almost always (very happily married group)<br>35–41% always or almost always (all other groups)<br>1–9% rarely<br>3.2–19% never |
| Gebhard et al. (1970) | 1,883 | 59% always or almost always (very happily married group)<br>35% always or almost always (unhappily married group)<br>10% no orgasm with coitus, although women are orgasmic |
| Hamilton (1929) | 100 | Note: Number of respondents is listed since it's the percentage given $N = 100$. Participants in this study were all members of heterosexual couples and both the men and the women were surveyed concerning the occurrence of female orgasm with intercourse. Female responses regarding how often they achieved orgasm are given first. Male responses concerning how often their partner achieved orgasm are then listed in parentheses.<br>She always has an orgasm: 6 (13)<br>Almost always; in 90% or more of sex acts has orgasm: 20 (14)<br>Usually has it: 7 (9)<br>Has it 75% to 89% of sex acts: 5 (4)<br>Has it 50% to 74% of sex acts: 5 (4) |

TABLE 1 *(continued)*

| Researchers | Sample size | Results |
|---|---|---|
| Hamilton (1929), con't. | | Has the orgasm but "not every time": 2 (1) |
| | | Has it 20% to 49% of sex acts: 5 (9) |
| | | Has it "frequently": 0 (1) |
| | | It is variable: periods when she has an orgasm every sex act and periods where she has no orgasms: 4 (1) |
| | | "Infrequently," "seldom," etc.: 4 (11) |
| | | Has had the orgasm only a few times in all her life: 2 (3) |
| | | Has had only 1–3 orgasms in all her life: 4 (1) |
| | | Is doubtful if she ever had an orgasm: 11 (5) |
| | | Never had an orgasm: 20 (10) |
| | | Has orgasms only by masturbation—never in the sex act: 0 (2) |
| | | Has multiple orgasms: 5 (5) |
| | | "Don't know" and other inconclusive answers: 0 (7) |
| Heyn (1921) | 512 | 51% always |
| | | 31% most of the time |
| | | 17% never |
| Hite (1976) | approx. 3,000 | Total population: |
| | | 26% regularly from intercourse |
| | | 19% rarely from intercourse |
| | | 16% orgasm with clitoral stimulation by hand |
| | | 24% no orgasm from intercourse |
| | | 12% never has orgasm |
| | | 3% never has intercourse |

TABLE 1 *(continued)*

| Researchers | Sample size | Results |
| --- | --- | --- |
| Hite (1976), con't. | | Population who does have orgasm and has had intercourse:<br>30% regularly from intercourse<br>22% rarely from intercourse<br>19% orgasm with clitoral stimulation by hand<br>29% no orgasm during intercourse |
| Hunt (1974) | approx. 700 | 53% all or almost all the time<br>7% almost none or none of the time |
| Kinsey et al. (1953) | 5,940 | 39–47% always or almost always<br>11–15% usually or regularly<br>24–30% sometimes or rarely<br>11–33% never |
| Kopp (1934) | 8,544 | 34% usually/always<br>46% occasionally/seldom<br>17.6% never<br>2.3% not sure |
| Landis et al. (1940) | 44 | This work reports "sex adjustment" in marriage for 44 "normal" women<br>. . .never experienced orgasm. . .: 7<br>. . .may never achieve orgasm. . .: 23<br>. . .sometimes orgasm but not most of the time. . .: 32<br>. . .orgasm most of the time. . .: 38 |
| Laumann et al. (1994) | 1,610[a] | 28.6% always attain orgasm with partner (don't say how)[b] |

**TABLE 1** *(continued)*

| Researchers | Sample size | Results |
|---|---|---|
| Levine and Yost (1976) | 59 | 45% attain orgasm 100%<br>2% attain orgasm 80%<br>2% attain orgasm 75%<br>10% attain orgasm 67%<br>12% attain orgasm 50%<br>12% attain orgasm 33%<br>17% no orgasm |
| Loos et al. (1987) | 87 | 7% always<br>24% nearly always<br>22% frequently<br>15% occasionally<br>16% rarely<br>11% never |
| Raboch and Bartak (1983) | 1,652 | 66% mostly or always<br>24% rarely<br>10% never |
| Raboch and Raboch (1992) | 2,425 | 52.2% always or mostly<br>25.4% infrequent or rarely |
| Rosenthal (1951) | nss[c] | 65% usually or always<br>12% sometimes<br>23% never |
|  | nss[d] | 27% usually or always<br>30% sometimes<br>43% never |
|  | nss[e] | 67% usually or always<br>25% sometimes<br>8% never |
| Schnabl (1980) | 4,000 | 26% always<br>17% "in most cases"<br>12% "often"<br>19% "sometimes"<br>17% "very rarely"<br>9% never with intercourse |

TABLE 1 *(continued)*

| Researchers | Sample size | Results |
|---|---|---|
| Slater and Woodside (1951) | 200 | Always: 30%<br>Often enough: 19%<br>Insufficiently/infrequently: 36%<br>Doubtful/never: 5%<br>No information: 10% |
| Stanley (1995) | 3,450 | 18% always<br>42% nearly always<br>11% sometimes<br>21% rarely or never<br>8% misc. replies |
| Stone and Stone (1952) | 3,000 | 41% regularly<br>43% "occasionally or rarely"<br>16% never |
| Tavris and Sadd (1977) | 100,000 | 15% always<br>48% most of the time<br>19% sometimes<br>11% "once in a while"<br>7% never |
| Terman (1938) | 760 | 22.1% always<br>44.5% usually<br>25.1% sometimes<br>8.3% never |
| Terman (1951) | 556 | 23% always<br>47% usually<br>25% sometimes<br>5% never |
| Wallin (1960) | 540 | 23% always<br>49% usually<br>28% never or sometimes (never = 5%) |
| Woodside (1950) | 48 | Always: 46%<br>Often: 29%<br>Infrequently: 21% |

TABLE 1 *(continued)*

| Researchers | Sample size | Results |
|---|---|---|
| Woodside (1950), con't. | | Never: 2%<br>No information: 2% |
| Yarros (1933) | 174 | 27% highly satisfactory<br>24% fair reaction<br>37% poor reaction<br>12% hate the whole thing |

a. This stratified sample is discussed later in the text.

b. Their survey question asked whether the woman had orgasm always, usually, sometimes, rarely, or never, but Laumann et al. report only the "always" rate. Also, the males reported the female orgasm rate of "always" at 40.2%, considerably higher than the female-reported rate of 28.6%.

c. For a population of British women, mainly middle class. No sample size is available for this figure.

d. For a population of French Belgian women, no sample size available.

e. For a population of American women, no sample size available.

Hence, even in the best-designed studies, the rate at which women "always" have orgasm with intercourse is quite low in the United States and the United Kingdom.

The numbers pick up quite a bit if the category is "always or almost always," which 6 studies researched. This is a category used by Kinsey and colleagues (1953) and Gebhard (1966, 1970) to indicate women who have orgasm from intercourse 90–100% of the time. The other 6 authors in this category may well have used different percentages. The results ranged from 11% to 60%, with two 59% figures coming from small subsamples of Gebhard's two studies, consisting of women who described themselves as "very happily married." The mean of the results was 41%, with a median of 43%.[8]

A more inclusive category consists of women who "usually, regularly, or always" have orgasm with intercourse. I estimate this category to include women who have orgasm at least 50% of the time with intercourse. Of the 25 studies from which these figures appear or can be calculated,[9] the percentage of women who fit into it ranges from 23% to 75% with a mean of 55.4% and a median of 55%.

Lower rates of orgasm with intercourse are represented by the "sometimes" and the "rarely" categories. Of the 15 studies that gave figures for "sometimes," 5% to 32% of respondents report that they sometimes have orgasm with intercourse; the mean is 19.7% and the median is 24%.[10] I should note the wide variety of women's responses that are included in this category. Some authors, such as Schnabl, differentiate the "sometimes" category from the "rarely" category, while others, such as Butler, do not. In addition, 2 studies gave figures for "sometimes or never" and their numbers were 58% (Clifford) and 60% (Fisher), while a third (Wallin) gave a figure for "up to half the time" of 28%. Among the 14 studies that included a separate "rarely" or "infrequently" category, there is a range of 1% to 46%, with a mean of 19.6% and a median of 17.5%.[11] Two authors have a category of "never or rarely," and their values are 7% and 21%, with a mean of 14%.

A total of 27 studies give rates for women who *never* have orgasm with intercourse. The range is 2% to 43%, while the mean is 12.4% and the median 11%. In many of these cases it is unclear whether the researchers are including women who never have orgasm under any circumstances, or only those who are orgasmic but not with intercourse. In the case of Hite, her 24% explicitly excludes those women who are not orgasmic or who do not have intercourse. Taking only the orgasmic women who have had intercourse as the population, Hite reports a frequency of 29% who do not have orgasm from intercourse. Kinsey's range of 11% to 33% of women who never have orgasm with intercourse certainly includes orgasmic women. The rates vary according to how many years the women have been married. Moreover, Gebhard, one of the coauthors of Kinsey et al. 1953, explicitly includes only orgasmic women in his 10% rate of women who don't have orgasm with intercourse (1966, p. 91). As for the rest of the studies, we simply do not know what they were counting.

In sum, it seems that approximately 25% of women always have orgasm with intercourse, while a narrow majority of women have orgasm with intercourse more than half the time. From the studies reported, roughly one third of women rarely or never have orgasm with intercourse, while approximately 23% "sometimes" do. We have seen that there is a fair degree of variation in the studies in the middle ranges of orgasm frequency, but better agreement on the "always" end and the "never" end. In nearly every study, there are serious problems with whether they are representative of any broader population, and they also suffer from being exclusively either North American or European. Still, given the information that has been collected over a period of decades, it is clear that one cannot assume, by any stretch of the imagination, that intercourse yields orgasm always or almost always for more than a minority of women. These results will be revisited in my consideration of various evolutionary accounts of female orgasm. They can be summarized as composite means of values from Table 1: 25% always have orgasm with intercourse; 55% have orgasm with intercourse more than half the time; 23% sometimes have orgasm with intercourse; 33% rarely or never have orgasm with intercourse; 5–10% never have orgasm at all.

It is also noteworthy that in several studies, for example Terman (1938) and (1951), the lack of orgasm with intercourse was treated as a lack of orgasmic capability altogether. Most studies of orgasm frequency did not include a separate category for women who had never had an orgasm at all; they seem to have been lumped in with women who did not attain orgasm with intercourse.

There are data on the frequency of women who never have orgasm at all. Kinsey and colleagues (1953, p. 513) found a frequency of 9% in his total population of 5,940 women. Other studies reported a range of figures: Levine and Yost (1976) found 5% of their

59 women; Fisher (1973) found 5–6% of his 285 women; while Landis and colleagues (1940) found 3% in a sample of 44. Some studies reported higher figures. Hite found 12% out of her sample of approximately 1,800 (1976, p. 230), while Christensen and Hertoft (1980) reported a figure of 20% of their sample of 32, although there is some reason to think that this number indicates the percentage of women who never had orgasm with intercourse, and not the figure on orgasmic capacity itself. Gebhard and colleagues (1970, p. 30) noted that their 10% rate of 1,883 women who had no orgasm with intercourse did not represent women who could have no orgasm at all, since it included women who had orgasm in petting or masturbation. Their total figure for women who were never orgasmic at all is 7% (1970, p. 55). Thus most of the numbers on the frequency of women who never have orgasm range between 5% to 10%.

In sum, if we look at masturbation, we get a very different picture of female sexual response from the one available from studies focusing only on orgasm with intercourse. In the masturbation studies we find more easily and quickly orgasmic women. The contrast between the two portraits of female sexuality led many sex researchers, most preeminently Kinsey and colleagues and Masters and Johnson, as well as Fisher and others, to conclude that intercourse often does not provide the right kind of sustained stimulation of the clitoral area to induce female orgasm. As Kinsey wrote, "It is true that the average female responds more slowly than the average male in coitus, but this seems to be due to the ineffectiveness of the usual coital techniques" (1953, p. 164). In addition, he wrote that "the techniques of masturbation and of petting are more specifically calculated to effect orgasm than the techniques of coitus itself" (1953, p. 391). Ford and Beach noted that "there is some question as to whether clitoral stimulation is usually necessary for complete climax. Certainly this is not the case for every human female. Nevertheless, there is no doubt that for a large proportion of women the clitoris serves as one important locus of stimulation" (1951, p. 22).

These contrasting pictures of female orgasmic capacity also serve as the basis for doubts that female sexuality should be viewed through the lens of intercourse, as has been done in the majority of studies to date. As Hite put it, "To assume that intercourse is the basic expression of female sexuality, during which women should orgasm, and then to analyze women's 'responses' to intercourse—is to look at the issue backwards" (1976, p. 60). Instead, it makes more sense to look at when and how women have orgasm first, and to view their sexuality in those terms. Whether and when they have orgasm with intercourse then becomes a specialized sort of question about female sexual response, and not its defining term. Under this approach, which is taken by some of the leading sex researchers, female sexuality is seen as independent of intercourse, and its expression in intercourse is seen as a (sometimes problematic) subset of overall female sexual response. The autonomy of female sexual response from intercourse turns out to be a major issue in evaluating many of the evolutionary accounts I shall examine.

Before continuing, let me emphasize one point. All sexual activities discussed here are frequently associated with female sexual excitement—with physiological changes such as vaginal lubrication and the engorgement of the vagina, clitoris, and surrounding tissues with blood, as well as the more general increases in heart rate and respiration. Masters and Johnson (1966) describe these changes in great detail. Their account has been criticized for not clearly delineating between the "excitement" phase and the "plateau" phase (Robinson 1976); let me here include the plateau phase in what I am calling "sexual excitement." Thus a woman is often aroused through foreplay, which puts her into a state of sexual excitement in which her vagina is lubricated, which in turn makes intercourse easier and more pleasant for both the man and the woman. But this state of sexual excitement is quite distinct from the existence of orgasm; excitement is necessary for orgasm, but it is not sufficient. Many women, for example, enjoy intercourse and find it exciting, but do not have orgasm from it. Women also enjoy engaging in intercourse for the feelings of closeness and intimacy with the male partner. The clitoris plays a crucial role in inducing sexual excitement, which in turn facilitates reproductive sex (intercourse). Thus the clitoris may very well have been under selection pressure to perform the function of inducing sexual excitement, thereby increasing female sexual interest in and capacity for intercourse. But note that the evolutionary account of female sexual excitement is distinct from an account that concerns orgasm, since sexual excitement often occurs in the absence of orgasm. Orgasm is a reflex that requires sexual excitement, but nothing about sexual excitement implies that orgasm will take place. This fact is most clearly seen when we consider nonhuman primate females, who experience physiological changes interpreted as sexual excitement, but who rarely experience orgasm (see the discussion in Chapters 3 and 4).

### Methodological Issues and Evidence

There are many methodological problems with some of the research I have been discussing. The most serious concerns the question of whether the surveys actually represent any sort of accurate portrait of the population of women as a whole. Most of them involve small, specialized subsamples of the population; it is unknown how the specialized nature of these samples might distort the picture the studies purport to paint about the female population at large. Does having the sample represent only upper-middle-class individuals make a difference? Kinsey and colleagues' results indicate that it does (1948, 1953). What about having only white women in the sample? What difference does it make that some of the samples were gathered from sex clinics, where the participants were there only because of some sexual problem? What about including only gynecology patients? Wouldn't this represent a specialized subsample of the population of women? The problem is that we can imagine all sorts of ways that these factors might make a difference, but without a truly representative sample, we have no idea how to correct for the biases (Cochran et al. 1953; Streitfeld 1988).

As I have noted, however, there were two studies cited that claimed to be genuinely representative of the population as a whole: the Mass-Observation study from 1949 (analyzed in Stanley 1995) and the Laumann et al. study from 1994. The first study turned out to be only partially a random sample, while the second study has come under fierce attack for its claims to represent the population as a whole. The Laumann et al. study involved 3,432 people, and its authors claim that this represents the nation as a whole. The authors originally picked a random sample of 9,000 people drawn from census data. Of these leads, many were unusable because no one lived at the specified address, there were no English speakers there, or there was no one between the ages of 18 and 59. The remaining sample was then supplemented with extra black and Hispanic people, because they were thought to be underrepresented. This is called a "stratified sample," and it is an accepted form of probability sampling. The problem with using this method is that the authors must decide which form of classifying groups is the important one, which groups to make sure are not underrepresented. Religion, for example, was not considered an important enough variable to make sure that there was adequate sampling. Thus nontrivial sociological decisions about what factors were relevant were built into the "random" sampling method itself. Moreover, as Richard Lewontin argues, further such loaded decisions were made regarding which information was considered salient (2001, pp. 254–256). Social class, for example, a variable that played a major differentiating role in the Kinsey studies, was not considered separately in Laumann and colleagues' analysis.

Lewontin is also critical of the Laumann et al. study's representativeness because many of the people approached for the study refused to cooperate. This may produce a bias in the survey results. "Are people who refuse to cooperate with sex surveys more prudish than others, and therefore more conservative than the population at large in their practices? Or are they more outrageous, yet sensitive to social disapprobation?" (2001, p. 258). There is no way of knowing how serious and in what direction the nonresponse bias is. The researchers did end up getting a 79% response rate, through being persistent and offering money, but it remains unknown how the exclusion of the other 21% skewed the results.

These are standard problems with random sampling methods, and they raise doubts about how representative the Laumann et al. sample really is. But there is a further problem, not confined to Laumann et al.'s study, and that is the problem of self-reporting. Because sex usually occurs in private, we have little public access to people's sex lives. Hence, we can almost never directly observe or study people's actual behavior. We are instead dependent upon what they *tell* us they do. Kinsey and his team were acutely aware of the problems of self-reporting. They knew that people often, sometimes even unconsciously, try to tell the interviewer something that would put them in a good light; that they will often self-report inconsistently; and that they can be strongly affected in their answers by the subtle attitudes and biases of the interviewer. As a result, Kinsey developed an extremely sophisticated interviewing technique, involving cross-checks on answers, a morally neutral and open demeanor of the interviewer, and careful controls for deception (Kinsey et al. 1948, pp. 35–150; Kinsey et al. 1953, pp. 22–97; Robinson 1976, pp. 44–48). The interviewers on Kinsey's project underwent extensive three-month training before they were allowed to collect field evidence. The situation with the Laumann et al. study seems to be quite different. Their interviewers underwent a three-*day* training session, and then were sent out into the field to collect data. Fewer cross-checks on false or exaggerated answers were used. Essentially, a stranger interviewed people for one and a half hours regarding the most intimate details of their lives. Moreover, some questions were considered more delicate, and were separated off into a paper questionnaire that went along with the interview. Lewontin expresses incredulity that such an approach could yield answers that were not affected by the interviewees' desire to impress and please the interviewer. He cites as an example the average number of heterosexual sex partners men and women claimed to have had in the past 5 years, pointing out that, analytically, the average number reported by men must be equal to the average number reported by women (2001, p. 262). What Laumann and colleagues found was that men reported approximately 75% more partners than women. Laumann and colleagues admit that the most likely explanation of this is that "either men may exaggerate or women may understate." "So," Lewontin concludes, "in the single case where one can actually test the truth, the investigators themselves think it most likely that people are telling themselves and others enormous lies" (2001, p. 263).

Furthermore, the fact that all of the survey results reviewed earlier in this chapter are based on self-reporting, either through interviews or through questionnaires, probably biases the results toward reported rates of orgasm higher than those experienced. Ever since Freud, there has been a heavily normative equation drawn between a woman having orgasms with intercourse and her true womanliness and femininity, thus producing great pressure on women to have orgasm with intercourse. Given this enormous social pressure, the surveys are most likely to yield higher rates of orgasm than actually exist. However, this bias only further supports my main conclusion about these studies: women do not reliably have orgasm with intercourse. Similarly, masturbation rates may be underreported, owing to the social disapproval of the practice; and we have no reliable data about orgasm rates during homosexual encounters. In contrast, the physiological results regarding the mechanisms of excitement, orgasm, and reproduction are much cleaner, because they were obtained by objective measurement, rather than self-reporting.

Given the methodological problems just discussed with the sexology literature, what should our approach be to treating it as evidence? Simply put, we must use the evidence we have but without illusions about some of the studies' apparent flaws. First, no one study should be treated as representative of the population at large. We should instead look to trends in the studies taken as a whole for a more representative understanding. Second, we should be aware that face-to-face studies may artificially inflate reported rates of orgasm, especially of orgasm during intercourse. And third, we must bear in mind that almost none of the studies draw the crucial distinction between assisted and unassisted orgasm with intercourse. I shall insist, throughout the book, that the results reviewed in this chapter should be taken seriously as evidence that is relevant to evolutionary explanations of female orgasm. In one fashion, it is a simple matter of description; if a researcher is writing about the evolution of female orgasm, then he or she should use the best, most scientific description of the phenomenon being explained. The fact that the "best, most scientific description" may, in fact, have faults is not a good excuse not to use it. The majority of researchers reviewed in this book express this sentiment—they usually cite the sex research as evidence (whether it actually supports their view or not). Thus there is a professional recognition of the importance of appealing to the science of record—sex research—to support evolutionary claims about female orgasm. The primary problems arise, as we shall see, in the selective use of that evidence and in its misinterpretation. I shall begin to uncover these problems in Chapters 3 and 4.

CHAPTER 3

# Pair-Bond Accounts
# of Female Orgasm

Female orgasm is emotionally loaded; its importance to human social life and to personal sexual expression makes it difficult to evaluate in a way unaffected by nonscientific beliefs and biases. Consider the problems accompanying our understanding of the trait.
Compare, for example, human female orgasm with laughter. Each
of us has laughed many times, and we have probably seen thousands of others laugh in thousands of situations. We are thus in
a real sense ordinary experts about laughter, even the grouchiest
among us. We know how to recognize it, what situations it is likely
to occur in, the variety that there is in the ways people laugh, and so
on. The female orgasm, however, is different. Men, obviously, have
never experienced the phenomenon directly, and they are likely to
have experienced the phenomenon indirectly in only a very limited
number of women, and even then, communication about the phenomenon is overwhelmingly likely to have been affected by sexual
and social mores. Among women, one in 10 or 15 have not experienced the phenomenon, and women are likely to have observed female orgasm in a limited number of others, if any others at all. Discussion about others' experiences even among women is often only
conducted with close friends, and then perhaps not even openly.
Thus each of us has at best only a highly skewed, but emotionally
loaded, picture of the general phenomenon. This, of course, does

44

not prevent scientific study of the trait, but it does make it very important that scientific studies are used, rather than our intuitions, when we test hypotheses concerning female orgasm.

With this injunction firmly in mind, let us turn to the central issues of the next two chapters. I examine various adaptive explanations of the existence of female orgasm among human beings, challenging both their underlying and explicit assumptions in detail. I find the following problems among nearly all of the adaptive stories discussed in the next two chapters: researchers make assumptions about the nature and physiology of human female sexual response that contradict available clinical and survey evidence; they make generalizations about nonhuman primates that are contradicted by laboratory and field evidence; and they make assumptions about the nature of our hominid ancestors' social and sexual interactions without even an attempt at giving supporting evidence. These and other problems with the explanations in this chapter are summarized in Table 2, in the next chapter.

As promised in Chapter 1, I need to review the standards of evidence that I am applying in my analyses. I reviewed the general requirements for demonstrating that a trait is an adaptation there. Here I wish to emphasize the importance of assuming that those requirements are met. The *assumptions* at the heart of an evolutionary model, or explanation, play a particularly important role in the scientific adequacy of that model (Lloyd 1988/1994, chap. 8; Rose and Lauder 1996). To take the most obvious example, most evolutionary explanations of particular traits can safely assume the present *existence* of the trait in question; evolutionists are rarely in the business of making predictions about the existence of traits. So, evolutionary explanations of the small wing size of ostriches assume that ostriches do have small wing size for their weight. But many other assumptions can come into play as well. Evolutionary explanations of a trait always involve an assumption about what trait was possessed by the ancestors of the species in question. In this case, the evolutionary explanation consists of postulating certain selection pressures in the environments of the ancestors of ostriches, giving evidence for the existence of these selection pressures in the environment, and showing how these selection pressures would lead to smaller wing size (Sinervo and Basolo 1996). It is assumed here that ostrich ancestors had comparatively larger wings, more like those of other birds. One can, of course, give empirical evidence in the form of fossils and the comparison of shapes of organisms and the shapes and functions of their traits (comparative morphology and physiology) to support these assumptions.

There are yet other assumptions in evolutionary accounts that need to be supported with evidence. Perhaps the most obvious—but one that proves to be a serious problem for many evolutionary explanations of female orgasm—is that the trait being explained needs to be accurately described. If it is not—for example, if female orgasm is represented in a manner that contradicts the findings of contemporary studies of female sexuality—then the evolutionary explanation is undermined.

Thus when any of the assumptions involved in an evolutionary explanation are contradicted by available evidence, the explanation is undermined. For example, if it turned out that ostrich ancestors did not have larger wings than contemporary ostriches, that would challenge any explanation premised on large-winged ostrich ancestors. Most evolutionary explanations of human female orgasm incorporate assumptions about the ancestral mating system, about whether females tended to mate with one or more males, and under what conditions. These assumptions, too, are subject to contemporary empirical verification, although the verification can only be indirect, since we cannot actually observe which mating systems were used by our ancestors.

This point highlights the role of indirect evidence in evolutionary explanations generally; the central assumptions of *any* model concerning the evolution of a particular trait are frequently not testable by direct observations, with the exception of the direct correlation of trait and reproductive success, and the direct connection of the trait with a mechanism. We cannot directly observe the past selective pressures on a population over time—support for claims regarding past selective pressures must come indirectly from studies of the past geology, geography, and biology.[1] In addition, one may extrapolate from present selection processes to past ones (see Chapter 6). Similarly, we cannot directly observe traits that are not present in fossilized remains; this means that information regarding behavioral traits such as female orgasm must be inferred by other means.

There are three standard methods for inferring information about such ancestral traits (Lancaster 1975; Larson and Losos 1996). One is to examine and compare the traits of living species more and less closely related to the species whose evolution we wish to explain. The reasoning here is that closely related species more than distantly related species may resemble the ancestors that we hold in common with these species. This has made the study of the great apes and the Old World monkeys particularly important for understanding human evolution. A similar approach involves seeking out related species that now live in environments similar to our ancestral environments. The idea is that related species may respond in similar ways to similar environmental challenges. The other primary approach to making inferences about the traits of our ancestors is to examine contemporary human cultures that live in environments like those in which the trait in question is believed to have evolved. Thus there are general approaches, but no systematic method for supporting claims made about ancestral traits and ancestral selection pressures. The methods used depend on the trait under consideration.

In sum, an examination of any evolutionary account involves several key points. The first is to establish that there is such a trait and that it has a particular nature. Second, we must establish a connection between the genetics and the trait. Third, a link between genetic reproductive success and the trait must be established. Fourth, the mechanism linking the trait to fitness must be uncovered. Also, we need an historical account of how we arrived where we are, with regard to the trait. Finally, the assumptions in the account need to be tested against available or procurable evidence.[2]

Throughout the book, I shall ask whether or not the basic assumptions of various evolutionary explanations are supported by evidence. The explanations I examine are flawed with respect to such evidence in various ways. They may contradict known evidence, or they may involve assumptions that are unsupported but not known to be false. They may make assumptions that contradict those made in other evolutionary explanations, but there may at present be no good way to tell which assumption is likely to be accurate. They also may involve assumptions that are incompatible with known evidence, but the models themselves may be fixable with some minor alterations.

It may seem that, since the debates covered in this book involve whether or not female orgasm is an adaptation, that the topic is one within sociobiology, the adaptive study of social traits. In particular, the debate between Stephen Jay Gould and John Alcock (Chapter 6) seems to concern when to consider a trait to be an adaptation, which is a stock area of disagreement under the rubric of sociobiology. But I don't think approaching the female orgasm debates as a case study of sociobiology makes much sense. First of all, the first hypothesis concerning female orgasm as an adaptation appeared in 1966, well before the heated debates spurred by Edward O. Wilson in 1975. Second, I consider the trait of female orgasm to be a physiological trait or reflex, not a social trait. Finally, the debates themselves make virtually no appeal to the sociobiological literature. Thus I will consider the sociobiological debates no further in this book, and will focus on the adaptive and nonadaptive explanations of female orgasm. While it may be true that the Gould versus Alcock debate was heated up by views about sociobiology, the arguments stand on their merits, and should be considered as such.

But enough of generalities. Let's begin the survey of evolutionary explanations for female orgasm. The first set of explanations I consider revolves around the notion of a pair bond—an enduring monogamous partnership—between males and females. In Chapter 4 I consider non–pair-bond explanations and female-centered explanations. I shall start by examining one of the earliest and most well known examples of a pair-bond explanation.

### Desmond Morris's Account

Eleven of the 19 adaptive explanations of human female orgasm are based on the notion that a male-female pair bond is adaptive, and that female orgasm helped with pair bonding. By far the most famous and well developed of the pair-bond accounts is that of Desmond Morris (1967), and I shall begin by examining his theory, noting along the way different views taken by other authors. I shall also note problems with the various theories and their assumptions.

Morris was a field ethologist dabbling in human evolution, taking many of his ideas from the "Man the Hunter" hypothesis of human evolution. Although Morris's work was criticized by later researchers as being methodologically flawed and "panglossian" (Crook 1972, p. 253; Wilson 1975; though Crook also praises Morris, 1972, p. 249), it was nevertheless widely cited over a 20-year span and its basic premises were accepted or modified in other, more recent accounts.

Morris's adaptive explanation of female orgasm centers on the notion of the pair bond, an enduring attachment between a man and a woman. Morris attempts to establish that pair bonding would itself be an important evolutionary adaptation for our pre-hominid and hominid ancestors. Females would be prone to develop a pairing tendency because "males had to be sure that their females were going to be faithful to them when they left them alone to go hunting" (1967, p. 64). According to Morris, males would also be selected for their tendency to pair-bond. Since more cooperation was necessary for hunting, Morris claims, the dominant males had to "give" the weaker males more sexual rights; the females would not be hoarded by the dominant males, but would be shared (1967, p. 64; see also Crook 1972, p. 249; Wilson 1978, p. 141).

Other authors emphasize a variety of different reasons that the pair bond would have been an important adaptation for early hominid life. George Pugh, for example, cites the increased exposure resulting from the move from a forest to a savanna environment, which in turn might have increased the female's need to be protected by the male, as well as increasing the economic dependence of females on males with the advent of hunting (1977, p. 250). Pugh, along with Niles Newton, John Crook, and David Barash, also cites the increased period of infant dependency and the need to provide a secure social environment for rearing offspring as an important pressure that might lead to the selection of the pair bond (Crook 1972, p. 249; Newton 1973, p. 92; Barash 1977, p. 297; Pugh 1977, p. 250). It may well be true that having a biparental family would be a benefit to the offspring. However, for reasons mentioned below, it seems implausible that female orgasm would provide the reason for the parents to stay together. Nevertheless, there may be other motivations, unrelated to female orgasm, although no such explanation has been offered. A third reason the pair bond might be selected is put forward by Frank Beach. On his view, the lack of certainty about fertility due to loss of estrus necessitated an increase in frequency of copulation.[3] Beach believed that having estrus was the ancestral trait for hominids. Therefore, he claims, the pair bond developed in order to make frequent and regular heterosexual coitus more efficient (1973, p. 360).

Thus although there are a variety of reasons presented, all of these authors saw the pair bond as a development in the course of hominid evolution that provided crucial selective advantages. (See Donna Haraway's summary of arguments against the scientific acceptability of the pair bond; 1991, chap. 5.) The next step in the argument involves linking the pair bond to the evolution of female orgasm. Once again, there is a range of arguments for such a linkage, but I shall begin with Morris's, both because it is the most complete and because it is the most problematic.

Taking himself to have established the evolutionary importance of the pair bond, Morris gives a surprising explanation of the evolutionary role of intercourse. Intercourse is supposed to be a way to maintain the pair bond: "the vast bulk of copulation in our species is obviously concerned not with producing offspring, but with cementing the pair-bond by providing mutual rewards for the sexual partners" (1967, p. 65). Such rewards are supposed to include orgasms for the females. Thus because pair bonds are adaptive for supporting the social structure of hunting, anything that would help maintain the pair bond is also adaptive, including female orgasm. Note that Morris's account works only if there is a differential rate of female orgasm within and outside of the pair bond.

Morris introduces his adaptive story about female orgasm with a summary of human sexual response. He seems to generalize from the male response to the female while ignoring the available information about female sexuality. Morris is aware that the frequency of female orgasm in intercourse is lower than the male's, but he claims: "If the male continues to copulate for a longer period of time, the female also eventually reaches a consummatory moment . . . some females may reach this point very quickly, others not at all, but on the average it is attained between ten and twenty minutes after the start of copulation" (1967, p. 54). He observes, "It is strange that there is this discrepancy between the male and female as regards the time taken to reach sexual climax and relief from tension" (1967, p. 55).

One problem with Morris's description is that it turns out not to be true that females take longer than males to reach orgasm; that only happens during intercourse. Recall that Kinsey and others have found that the time to orgasm in women (during masturbation) is the same as for men (Kinsey et al. 1953, p. 164).

Morris continues, "After both partners have experienced orgasm [in intercourse] there normally follows a considerable period of exhaustion, relaxation, rest and frequently sleep" (1967, p. 55). Similarly, he later asserts, "once the climax has been reached, all the [physiological] changes noted are rapidly reversed and the resting, post-sexual individual quickly returns to the normal quiescent physiological state" (1967, p. 59).

The tendencies to states of sleepiness and exhaustion are, in fact, predominantly true for men but not for women (Kaplan 1974, p. 31). Regarding Morris's claim that the physiological changes are "rapidly reversed," that is also true for men but not for women, who return to the plateau phase of excitement, and not to the original unexcited phase (Masters and Johnson 1966, pp. 283–284). This is supported by the experiences of the women in Shere Hite's study, almost all of whom did not indicate a return to the base state. They described two basic kinds of feelings—wanting to be close, and "feeling strong and wide awake, energetic and alive." Hite remarks that "both of these reactions represent continued arousal" (1976, pp. 86, 87). David Shope reports that 25% of women who had orgasm with coitus reported increased energy after coitus during daylight hours (1968, p. 213). We must remember Masters and Johnson's physiological description of the basis for continued arousal. As we shall see, Morris's disregard of this important difference between male and female sexuality tells against part of his adaptive explanation.

Although Morris offers a quite technical description of female sexual response during intercourse, his account includes some odd features. Morris notes that human males have "the largest erect penis of any living primate" (1967, p. 58). According to Morris, the extra thickness of the penis is supposed to help stimulate the woman; it results in "the female's external genitals being subjected to much more pulling and pushing during the performance of pelvic thrusts." Add this stimulation to the pressure from the pubic region of the male, says Morris, "and you have a repeated massaging of the clitoris that—were she a male—would virtually be masturbatory" (1967, p. 80).

Morris is clearly using Masters and Johnson's model of indirect clitoral stimulation caused by thrusting: "as the penis moves back and forth, it pulls the labia minora, which are attached to the skin covering the clitoris (the hood), back and forth with it, so indirectly moving the skin around over the clitoral glans" (Hite 1976, p. 272). In other words, the clitoris is supposed to be stimulated by friction against its own hood. But this model begins to look suspicious when we note that Masters and Johnson developed the model by observing the women in their study, all of whom were selected *only if they could consistently and easily have orgasms during unassisted intercourse.* According to the general statistics, cited in Chapter 2, such women represent perhaps 20% of the adult female population, and thus cannot be considered representative.

Morris's "virtual masturbation" comment quoted above is also curious, because, were she masturbating herself, unless she is among the very small percentage who masturbates by vaginal insertion alone, she would stimulate her clitoris and vulva *directly* (Kinsey et al. 1953, p. 158). There is a problem that Morris does not address given his characterizations of female sexual response, and it is very simple: If this indirect stimulation mechanism works so well, and is "virtually masturbatory," then why the low frequency of female orgasm with unassisted intercourse? And why don't women masturbate in a way that imitates intercourse? As Wendy Faulkner points out, "There is no logical reason for insisting that our orgasms result 'naturally' from intercourse" (1980, p. 151). I shall discuss this point again later on in this chapter.

So far, then, we have Morris's work conflicting with the available sex research in significant ways. First, he misrepresents the sex literature about timing to orgasm. Second, he assumes in several places that female response is like male response to the same situation, in particular to the friction from intercourse, and he neglects the available work on female masturbation.

One of the most fundamental aspects of Morris's view is his claim that nonhuman female primates do not have orgasms (1967, pp. 54, 63). That is why human female orgasm needs a unique explanation in terms of the hominid lines. He explains that, in nonhuman primates, female orgasms would "waste valuable potential mating time" during estrus; it would be more adaptive if instead the females spent their time having more copulations to ensure successful fertilization (1967, p. 78). This seems far-fetched.[4] A rhesus macaque in estrus may copulate with 20 males per day (Goy and Goldfoot 1975, p. 418). If she were to take perhaps a little more effort per encounter so she could have her own orgasm, it is not at all obvious that this would substantially lower her chances of getting pregnant. Perhaps it would, but Morris needs evidence to make this claim.

But the real problem here is that there is robust evidence—developed since Morris wrote—that some nonhuman primate females *do* have orgasm. The best evidence comes from experiments in which stumptail macaques were wired up so that their heart and respiration rates and the muscle contractions in their uteruses or vaginas could be measured electronically (see Chapter 5; Goldfoot et al. 1975, 1980; Slob et al. 1986; Slob and van der Werff ten Bosch 1991). Researchers were prompted to wire up the stumptails because of previous observations by Suzanne Chevalier-Skolnikoff (1974, 1976) that showed a "naturally occurring complete orgasmic behavioral pattern for female stumptails." She documented three occasions on which a female mounting another female (rubbing her genitals against the back of the mounted female) displayed all the behavioral manifestations of male stumptail orgasm and ejaculation (1974, p. 109).

There were several intriguing results in the electronic studies of female orgasm in stumptail macaques. First, there was "no clear relation of the [orgasmic] response to any particular phase of the [menstrual] cycle" (Goldfoot et al. 1980, p. 1478). Second, the incidence of female orgasm when one female was mounting another female was much higher than when a female engaged in heterosexual intercourse (Goldfoot et al. 1980, p. 1477). Moreover, in two studies, the researchers found clear electronically measured evidence of orgasm only when females mounted other females, but not when the females were engaged in heterosexual intercourse (Goldfoot et al. 1975; Slob and van der Werff ten Bosch 1991; note that this is my interpretation of the data in the second study, not theirs). Other evidence for the existence of female orgasm in chimpanzees and in rhesus macaques comes from experiments involving human manual stimulation (Burton 1971; Allen and Lemmon 1981).

I shall discuss the evidence for female orgasm in nonhuman primates further in Chapter 5. For now, the important points are that, contrary to the claims of Morris, there is now good evidence of female orgasm in some nonhuman primates, and also good evidence that the occurrence of orgasm seems to be dissociated from copulation. Although only anecdotal evidence was available to Morris at the time he wrote his book, it may yet be relevant to our contemporary evaluation of his view. Thus one key question for us is whether the assumption that female nonhuman primates lack orgasm plays an important role in his account. The answer is that it is central, because Morris wishes to explain how loss of estrus and the advent of the pair bond—a suite of behaviors unique to the hominid lineage according to him—set up a selection pressure for female orgasm.

We must also ask whether his mistake about nonhuman primates reveals other assumptions in Morris's thinking, and I believe it does. First, it reveals an assumption tying sexual behavior to hormonal status. Morris assumes that our prehominid ancestors had clearly demarcated periods of sexual interest. All this was changed, according to Morris, through the demands of the pair bond, which required frequent intercourse for its maintenance. Second, Morris makes an assumption tying intercourse and female orgasm. For example, Morris does not even consider the possibility that nonhuman primate females might have orgasms when not in estrus, although it turned out later that most observed orgasms in stumptails occurred while the monkeys were *not* in behavioral or hormonal estrus and indeed were mounting other females. Morris fails to consider this possibility both because he assumes that female monkeys experience sexual interest only when they can be made pregnant and because he assumes that intercourse is necessary to female orgasm. Much less could he have imagined the sexually active bonobos, who seem to have sex in the wild about two-thirds of their cycle.

Morris claims that in human beings, unlike nonhuman primates, there is "nothing working against the existence of female orgasm"; since there is only one male involved, he argues, "there is no particular advantage in the female being sexually responsive at the point where he is sexually spent" (1967, p. 78). Thus female orgasm is supposed to serve as a way to stop being responsive. Unfortunately for Morris's theory, even if female orgasm were reliably tied to intercourse, a woman's orgasm does not stop her from being sexually responsive or sexually aroused, as Masters and Johnson showed (1966). This fact is particularly damaging; suppose females got attached to having orgasms, were not satisfied by the usual brief copulation with one male, and sought out other sexual contacts to satisfy themselves—this would work against the pair bond, and on Morris's scenario, would be selected against. Note that, again, his argument depends on assuming that the conditions for sexual satisfaction for the male will also produce sexual satisfaction for the female. Morris is quite explicit about his assumption. Female orgasm evolved because it contributes to the pair bond through "the immense behavioral reward [including orgasm] it brings to the act of sexual cooperation with the mated partner" (1967, p. 78).

Morris also offers an explanation for female orgasm completely distinct from issues about pair bonds. He claims that female orgasm would be selected—it is an adaptation—because "it considerably increased the chances of fertilization . . . in a rather special way that applies only to our own peculiar species" (1967, p. 78). Morris argues that "there is . . . a great advantage in any reaction that tends to keep the female horizontal when the male ejaculates and stops copulation [because that would keep the sperm from leaking out]. The violent response of female orgasm, leaving the female sexually satiated and exhausted, has precisely this effect" (1967, p. 79). As noted above, a sizable proportion of women are not "satiated and exhausted" by orgasm but, rather, energized and aroused. An "energized" woman seems less rather than more likely to lie down. Also, Morris assumes that the woman is lying on her back. He claims that the face-to-face posture is the "natural" one for human beings, that "the frontal approach provides the maximum possibility for simulation of the female's clitoris during the pelvic thrusting of the male," and that having the male on top of the female is the "most *efficient* and commonly used position" (1967, p. 74; my emphasis). This last claim is wrong. The most efficient position for the stimulation of the clitoris during intercourse is with the woman on top of the man (Ford and Beach 1951, p. 23; Masters and Johnson 1966, p. 59). Given that a relatively low percentage of women have orgasms during intercourse, and that of those who *do*, a high percentage have them in the superior position, it seems more likely that the occurrence of female orgasm would have the reverse gravitational effect from the one that Morris describes (Kinsey et al. 1953, pp. 362–364; see Kinsey et al. 1953, p. 363, for anecdotal evidence for the greater likelihood of female orgasm in the female-superior position; note that the female-superior position provides greater clitoral stimulation, as documented by Masters and Johnson 1966, thus providing the sort of stimulation helpful to female orgasm).

This antigravity theory of Morris's is worth dwelling on for a moment, because it has been influential. Gordon Gallup and Susan Suarez agree with Morris that bipedalism poses a unique and "formidable" reproductive problem: Gallup and Suarez claim that the resumption of an upright posture following copulation "would clearly endanger sperm retention . . . one would expect to find behavioral adaptations which would serve to compensate for, or at least postpone this eventuality" (1983, p. 195). Thus they offer their own main hypothesis for the evolution of female orgasm: "it is widely acknowledged that intercourse frequently acts as a mild sedative. The average *individual* requires about five minutes of repose before returning to a normal state after orgasm, and some people even lose consciousness at the point of orgasm (Kinsey et al. 1948)" (Gallup and Suarez 1983, p. 195; my emphasis).

There could not be a better example of researchers failing to differentiate between female and male sexual response than Gallup and Suarez; in this explanation of female orgasm their "average individual" is drawn from Kinsey et al. 1948, a book that is, in fact, about *male* sexual response only, and that contains no data about female sexual response. Although it is true that Kinsey et al. emphasized the similarities between male and female response, later studies, such as those done by Masters and Johnson, found significant differences in male and female responses to orgasm, notably including the female's continued state of arousal after orgasm (1966, p. 283).

Curiously, in Morris's discussion there emerges a third entirely different kind of idea about the origin of female orgasm. After noting that the female orgasmic pattern is almost identical with that of males, Morris suggests that "perhaps it is in an evolutionary sense a 'pseudo-male' response," a latent property in females acted upon by evolution (1967, p. 79). He reminds us that the clitoris, which in our species has developed a particular susceptibility to sexual stimulation, arises from the same embryological tissue as the penis (1967, p. 79). I shall discuss this theory in Chapter 5.

In this discussion of Morris's two main explanations for the evolution of female orgasm—the pair-bond account and the antigravity account—I have detailed various points of tension between Morris's depiction of female orgasm and the results from studies of female sexual response. Given that Morris seems to have repeatedly made claims about female sexual response that differ from those which researchers had documented and have since documented about female sexuality, I conclude that there are serious flaws in Morris's description of the trait he is trying to explain. This is especially surprising given that he cites some of the very studies—for example, Kinsey et al. (1953), Masters and Johnson (1966)—that undermine his own account. As mentioned earlier in this chapter, faulty descriptions of the trait being explained may be a sufficient reason to reject an evolutionary account. In Morris's case, such misrepresentations do affect the legitimacy of his account. For instance, his assumption that female orgasm occurs reliably with intercourse is central to his explanation of the evolution of orgasm in relation to the pair bond. Without the faulty description, his story does not get off the ground.

Morris has a tendency to assume that females respond just like males to the occurrence of intercourse and to the occurrence of orgasm, despite the evidence—readily available at the time from the Kinsey report he cited and numerous other sources—documenting the differences between male and female sexual response to intercourse. In other words, Morris ignores what I shall henceforth call "the orgasm/intercourse discrepancy."

Morris does acknowledge in one place that the male and female responses to intercourse are different, but he mischaracterizes the situation as one in which the fundamental female response simply takes longer than the male response. Most revealing, perhaps, is his surprise at the fact that males and females have different responses to intercourse, which he describes as "strange." It is clear at this point that he has a set of expectations about the "good fit" between male and female sexual responses during intercourse. My claim is that these expectations of "good adaptation" affected the rest of his account of the evolution of female orgasm.

My concern here is that Morris is simply assuming that female orgasm is an evolutionary adaptation, that it achieved its present form through direct selection on the trait. The fact that female orgasm has not been shown to be associated with any difference in reproductive success—in other words, we have no evidence that females with orgasm had relatively greater reproductive success than females without it—is a problem for such an approach. Thus Morris's assumption of good adaptation is unsupported. Yet despite its flaws, Morris's work gave rise to a number of other attempts to explain the evolution of female orgasm.

In fact, numerous early models were offered in the first wave of research on the evolution of female orgasm. I offer Table 2, in the next chapter, as a roadmap to the various hypotheses and their problems. A significant subset of these models focused on the role of female orgasm in promoting pair bonds, and it is to this group of models I now turn.

### Other Pair-Bond Accounts

It may be objected that Morris is an inappropriate target for criticism, because he was writing a speculative book for the lay public. However, the body of his explanation was advocated by a number of other authors who also appeal to some aspect of pair bonding in order to explain the evolution of female orgasm. Each of these accounts has its own idiosyncratic selective explanation, and they characterize female sexuality in various erroneous ways.

Pugh, for example, claims that sexual pair bonding was highly adaptive, and like Morris, sees the evolutionary problem as one of maintaining and encouraging that bond. Some of the adaptations in human beings that promoted sexual bonding, according to Pugh, included "many physical changes in the female designed to encourage a more permanent male-female relationship," including "the development of a female orgasm, which makes it easier for a female to be satisfied by one male, and which also operates psychologically to produce a stronger emotional bond in the female" (1977, p. 248).

Pugh leaves the claim that female orgasm "makes it easier for a female to be satisfied by one male" unsupported. In fact, as we have seen, the evidence regarding human female sexuality might undermine this assumption. The ability of females to have orgasms could do just the opposite of what Pugh says; given the relatively low rate of female orgasm with intercourse, intercourse often leaves the male but not the female satisfied, so it is difficult to see how the presence of female orgasm would incline a female toward being satisfied by one male. It seems that a *reduced* female sexual capacity would more efficiently accomplish this task.

John Crook, a serious and well-respected primatologist, ties the evolution of female orgasm to a variety of other features: "breasts of young women taken together with other features . . . seem to represent the main visual sexual releasers for the male . . . [and] have probably been selected in both sexes for their effect in improving sexual rewards, in inducing sexual love, and in maintaining pair bonds. The same is also likely to be true for the presence of orgasm in women" (1972, p. 254). Crook, praising Morris, claims that "through making sex sexier, as Morris puts it, the reward value of partnership in mating is increased and the pair bond maintained. Promiscuity is reduced through sexual love for one's mate and by the rivalry and aggression of other males already in possession of the other females" (1972, p. 250).

There are a number of problematic assumptions here. Like Morris, Crook assumes that females are satisfied sexually by one mate, and that males control and possess females. Crook does not address the problem of the relatively low frequency of female orgasm with intercourse. In addition, he assumes that continuous sexual receptivity is logically or evolutionarily linked to pair bonding. This assumption is undermined by the abundant primate data showing continuous receptivity in species that are not monogamous, as noted on pp. 64–66 of this chapter.

Newton's account is similar to Crook's and Morris's accounts, in that Newton emphasizes the role of female orgasm in cementing the pair bond. The twist is that Newton frames the case in terms of behaviorist psychology: "The intense emotions involved in orgasm are, in fact, a perfect model for operant conditioning. The pleasure male and female gain during coitus may tend to condition them to the other partner and to bind them in reproductive partnership so the children are more likely to have two adult individuals on the scene. *Operant conditioning, reinforced through coital pleasure, may be the biological function upon which patterns of family life are built*" (1973, p. 92; his emphasis).

In fact, Newton claims that there are specific domestic payoffs for the pair bond from female orgasm: "females who form satisfying mating relationships with men usually accompany the coital behavior with an urge to care for the man in various other ways—like cooking for him, making a home for him, and being emotionally committed to his well-being" (1973, p. 92). Thus Newton proposes a tight link between the existence of female orgasm and a feminine gender role itself. Note again the identification made between intercourse and the occurrence of female orgasm.

Irenaus Eibl-Eibesfeldt's explanation of female orgasm is similar in spirit to Newton's and the others I have discussed; it is one of the earliest accounts, and perhaps had some influence on later explanations. After explaining that the constant receptivity of the female throughout her menstrual cycle is an adaptation that "enables her

... to maintain a tie with the man on the basis of a sexual reward," Eibl-Eibesfeldt claims: "Also in the service of maintaining a bond between partners is that ability of the woman to experience an orgasm comparable to that of the man. This increases her readiness to submit and, in addition, strengthens her emotional bond to the partner" (1970, p. 443). This explanation runs afoul of the objections previously reviewed.

David Barash's twist on the story lies in his assertion that nonhuman animals copulate "almost exclusively with regard to reproduction." He continues: In humans, unlike other animals, "sexual activity is not limited to distinct breeding seasons or special periods of receptivity correlating with ovulation by the females." According to Barash, "humans are unique among animals in the additional use we make of sex" (1977, p. 296).

The puzzle, according to Barash, is to explain human beings' novel use of sex, sex's liberation from "its domination by hormones." He claims that this is explained by "the unique problem posed by our utterly dependent infants"; "given that, during our evolutionary development, offspring were more likely to be successful if they received the committed assistance of at least two adults, selection would favor any mechanism that kept the adults together." Barash continues: "sex may be such a device, selected to be pleasurable for its own sake, in addition to its procreative function. This would help to explain why the female orgasm seems to be unique to humans; among other animals, reproduction is the only goal, and satisfaction per se is irrelevant." He even goes so far as to claim that "sex for its own sake may in fact be one of the few biological distinctions of our species, a uniquely human attribute" (1977, p. 297). That he is talking about intercourse, and not about other forms of sexual activity is clear: "the enormous satisfaction of sex may be viewed as a trick by natural selection to ensure that we reproduce" (1977, p. 296). In other words, "sex" is reproductive sex—intercourse. So, for Barash, sex equals intercourse, which equals female orgasm, which equals an incentive to stick with a male partner to raise a child.

One serious problem with Barash's account is that it relies so heavily on the human uniqueness of continual female sexual activity. The nonhuman primate data undermine this cornerstone of Barash's account. Although it is widely believed that nonhuman primate females have their interest in sexual activity firmly circumscribed by their hormonal condition, this belief is contradicted by evidence from a number of species. Probably the most famous are our closest relatives, the bonobo and the chimpanzee. Bonobo females, in particular, have demonstrated sexual interest and participation outside their extremely extended period of hormonally influenced genital swelling (Kano 1980, p. 253; 1982, p. 186; 1992; de Waal 1995, p. 37). Chimpanzee females, too, have been observed having sexual interactions when they are outside their hormonally controlled estrus period—these observations date back to the beginnings of primatology (see the summary in Dixson 1998, chap. 5; Yerkes 1939a; Lemmon and Allen 1978; Goodall 1986, p. 445). As W. B. Lemmon and M. L. Allen put it, female chimpanzees are "receptive to copulation at virtually any time during adult life" (1978, p. 84). This finding might be an artifact of Lemmon and Allen's observations being based on chimpanzees housed in zoos, but Dixson (1998) also finds a number of species willing to copulate outside of estrus.

Evidence of sexual behavior and copulation outside of the hormonally controlled estrus period has been documented for a variety of species, including rhesus macaques (Rowell 1963; Loy 1970; Hafez 1971; Hanby et al. 1971; Lindburg 1983); baboons (Saayman 1970, 1975; Hafez 1971); orangutans (Nadler 1980); and Sykes' monkey and the vervet monkey (Rowell 1972). There is also abundant evidence of copulation while the female is already pregnant in rhesus macaques, chimpanzees, gorillas, Sykes' monkeys, vervet monkeys, and *Macaca fuscata* (Rowell 1972, p. 83).

In fact, it has been widely accepted for many decades that many Old World monkeys and apes, especially the orangutans, do not show well-circumscribed periods of estrus, but will copulate throughout the menstrual cycle (Miller 1928; Zuckerman 1932; Yerkes 1939b, p. 79; Michael et al. 1967, p. 554; Dixson 1998).

Barash's position is further challenged by observations of homosexual behavior among nonhuman primate females that are especially useful because they show a further dissociation between female sexual arousal and both heterosexual copulation and the estrus period. In heterosexual behavior, the results depend on two confounded factors: whether the male is interested and whether the female is sexually interested outside of her hormonally defined estrus period. The timing of the activity of the female who mounts other females is independent of male sexual response. In fact, in one captive group of rhesus macaques, the mounter's sexual activity was highest in the menstrual phase *before* the fertile ovulatory period—before the sexual swellings appeared and she entered the phase of high heterosexual activity (Michael et al. 1974, p. 406; Akers and Conaway 1979, p. 68). Stumptail macaque female mounters are also sexually assertive outside of the estrus period (Chevalier-Skolnikoff 1974, 1976). In addition, the hormone researcher Frank Beach (1968) gave evidence showing that the mountee (the mounted female) in homosexual interactions among nonhuman primates was usually in estrus, while the sexually active mounter was not. Beach later concluded, in his analysis of the relation between estrus and sexual activity, that ape and monkey copulation depends on individual affinities and aversions as much as on the presence or absence of hormones (1976, p. 130).

Hence, Barash's claims regarding the uniqueness of human or prehuman sexual activity throughout the cycle are uninformed by both the old and more recent literature on nonhuman primates. Moreover, Lemmon and Allen comment in a way that seems designed for the likes of the Morris and Barash theories: "the existence of continual female receptivity does not appear to be the over-riding factor of importance in the formation of the human pair bond any more than it accounts for primate social cohesion" (1977, p. 85).

In a different vein, Bernard Campbell's early evolutionary account provides an instructive example of the importance of distinguishing female sexual excitement from female orgasm. He begins by noting that "there is no doubt that in [nonhuman] animals the function of the clitoris is to raise the level of sexual excitement to the point necessary for effective intromission," and he mentions cases of nonhuman primate females stimulating their genitals by hand and orally. He claims that "orgasm appears to be an intensification of the sexual response seen in other mammals, brought about by increased clitoral stimulation and associated sensations" (1966, p. 260).

Thus Campbell claims that orgasm evolved to motivate the female to have intercourse: "it appears that in human evolution the loss of the innate and predetermined urges of estrus is, as it were, balanced by the more effective but casual stimulation of the clitoris that occurs not only before but during intromission" (1966, p. 260). This assumes two things: first, a strong link between the desire to have intercourse and the presence of estrus in our ancestors; and second, a link between female orgasm (which Campbell thinks of as uniquely human) and intercourse. Regarding the first assumption, Campbell must hold that our hominid ancestors, if they had lost estrus and had not evolved the orgasm, would not have been motivated to have intercourse. As we saw from the evidence from nonhuman primates, this is a problematic assumption, because the lack of estrus does not appear to bring about a lack of interest in intercourse in many of the monkeys and apes. Regarding the second assumption, like Morris, Campbell assumes that only human females have the capacity for orgasm and that this capacity is tied to a desire for intercourse (1966, p. 258). But we have seen that some female monkeys (for instance stumptail macaques) are, in fact, capable of having orgasm; and that those that are do *not* commonly experience orgasm during intercourse, but rather primarily while mounting other monkeys.[5] Hence, the nonhuman primate evidence documents both the presence of intercourse outside of estrus without orgasm and the frequent occurrence of orgasms outside of intercourse, thus doubly undermining Campbell's claim that the human female evolved the orgasmic capacity because she needed an incentive for intercourse once she lost estrus.

But perhaps our hominid ancestors were significantly different from these monkeys. In fact, Campbell believes that they were; more specifically, he offers an explanation for the fact that the clitoris is larger (relative to body size) in nonhuman primates than in human beings. The (supposed) human clitoris's small size, according to Campbell, is due to "a slight alteration in function due to a change in position in copulation. In particular, it is so effectively placed for stimulation during ventral copulation that excessive size is unnecessary and could be disadvantageous" (1966, p. 255). This conclusion would have come as a surprise to Kinsey and colleagues, who emphasized the inadequacy of the stimulation received by the clitoris during copulation (1953, p. 164). Strangely, Campbell actually *cites* Kinsey, but he overlooks Kinsey's conclusions. This neglect of Kinsey's conclusions was, I suspect, encouraged by Masters and Johnson's theory of indirect stimulation, which had just come out when Campbell was writing. In the absence of a better theory about how the ventral position stimulates the clitoris during intercourse, problems with the indirect-stimulation theory mentioned earlier apply equally well here.

Frank Beach, more than any of the other authors I consider here, emphasizes the significance of the loss of estrus for human survival. Having remarked that it is "taken for granted" that the primary reinforcement for sexual behavior in male mammals of all species is "the experience of orgasm which is assumed to accompany ejaculation," Beach notes that the rewards of copulation for the female "have not been dealt with by evolutionary theories in general" (1973, p. 358). Beach claims: "It is not necessary that women enjoy intercourse in order to become pregnant; but if pregnancy is to occur often enough to ensure continuity of the species, it is important that copulation be rewarding for both sexes . . . there is one source of reinforcement that can be directly tied in with changes occurring during human evolution": female orgasm (1973, p. 358). Hence, female orgasm evolved in order to encourage intercourse. Like Campbell, Beach believes that females of other species do not generally experience orgasm: "behavior indicating the occurrence of sexual climax in copulating females is extremely rare . . . the weight of available evidence favors the theory that female orgasm is a characteristic essentially restricted to our own species" (1973, p. 359). Notice that Beach, like others, has assumed that a lack of orgasm during copulation means the lack of orgasm altogether; we now know this inference to be unjustified, given the later nonhuman primate evidence reviewed above.

Beach links the evolution of the "unique" women's orgasm to several distinctly human anatomical and behavioral characteristics (1973, p. 359). For instance, he thinks that the female orgasm is an indirect evolutionary result of bipedalism among humans. Bipedalism resulted, he claims, in a repositioning of the female genitals further toward the front of the body, which "increased accessibility of the vagina to penile insertion" during face-to-face copulation. This face-to-face copulatory position increases "secondary or indirect stimulation of the clitoris," according to Beach, which in turn "increases the probability of feminine orgasm. Since capacity for orgasm is independent of the woman's current hormonal status, the possibility of its occurrence throughout her cycle tends to reinforce and increase the frequency with which she desires and accepts intercourse" (1973, pp. 359, 360). Beach supports his view by noting that, in many cultures, the superior position of the woman in coitus "is recognized as one method of maximizing her sexual pleasure and satisfaction" because of clitoral stimulation (1973, p. 359). First, the face-to-face copulatory position is widely but mistakenly believed to be unique to human beings. Bonobos, it turns out, copulate face to face almost as frequently as they do ventral-dorsally, and they are *not* bipedal (Mori 1984, p. 257). Since bonobos do have a more ventrally positioned vagina than common chimpanzees, we may provisionally conclude that the face-to-face position may be associated with a ventrally positioned vagina. But the argument that this vaginal position, face-to-face copulation, and in turn female orgasm are results of bipedalism is undermined.

Although Beach does show relative sensitivity to the facts of women's sexuality, there are two more problems with his explanation. First, he seems to assume early hominid or prehominid females' lack of receptivity and sexual interest during nonovulatory portions of their cycles, which is challenged by the nonhuman primate evidence just reviewed.

Second, on Beach's view, the lack of certainty about fertility due to loss of estrus necessitated an increase in frequency of copulation. Therefore, he claims, the family developed in order to make frequent and regular heterosexual coitus more efficient, and to reduce outside interference (1973, p. 360). He asserts that the family structure "would be likely to result in couples' copulation more often simply because of longer periods of proximity" (1973, p. 361). Kinsey and colleagues, however, provided extensive documentation of the fact that, the longer a couple is together, the *less* frequent their copulations (1953, p. 394). This is true despite the fact that, through the age of 35, the woman's sex drive and sexual interest actually increase. It seems more plausible to interpret the data on human sexuality as indicating that, if the woman is highly motivated to achieve orgasm through intercourse, she will be more likely to break up the pair bond, rather than reinforce it. Still, it is possible that ancestral populations behaved differently than did Kinsey's population. This is a weakness in the sexology literature that we have no means to counteract.

Finally, Beach is like the many researchers we have seen who seem unaware of the orgasm/intercourse discrepancy as a problem, since he claims that "copulation tends to result in mutual physical gratification" (1973, p. 361). But Beatrix Hamburg's account, which is closely tied to Beach's, and appeals to the same mechanism and cites Beach as her inspiration, is different in this respect. She addresses the orgasm/intercourse discrepancy directly (1978a, p. 163). Like Beach, she claims that in face-to-face copulation the anterior placement of the vagina "enormously facilitate[s] clitoral stimulation. It has been shown that female orgasm is chiefly dependent on rhythmic mechanical stimulation of the clitoris (Masters and Johnson 1966). The human female's ability to experience orgasm comparable to the male's *enhances the reward value for both* and gives sexual behavior an important role in interpersonal bonding" (1978b, p. 386; my emphasis). She might seem here to be ignoring the orgasm/intercourse discrepancy, but that is not so. She writes further: "The orgasm of the human female is very important and seems to be a fairly recent evolutionary acquisition. It has been sporadically reported in a few other species, but it clearly is not entrenched in the primate gene pool. Chimpanzees, for example, do not show this characteristic. Even among human females there appears to be a wide range of variability in orgiastic potency. This variability may also point to the relative recency of this trait's entry into the gene pool" (1978b, p. 386).[6]

Thus Hamburg acknowledges the orgasm/intercourse discrepancy, and offers the recency of evolution of the trait as an explanation of its variability. Because orgasm only recently occurred as an evolutionary variation or option, there has been insufficient time to establish it, through evolution by selection, as a robust and reliable response. In other words, female orgasm is seen as an adaptation to encourage women to engage in intercourse, thus aiding interpersonal bonding, but the reason it does not work that well is that the requisite variation has appeared too recently in evolutionary history to become fixed in the population. This explanation is an improvement upon the other pair-bond accounts, which do not even attempt to explain the orgasm/intercourse discrepancy. However, it also illustrates a potentially serious problem that is shared by all of the accounts in this chapter.

Specifically, Hamburg explains the evolution of female orgasm, full stop, by examining only orgasm with intercourse. As we know from Chapter 2, the capacity of women to have orgasm far exceeds their ability to do so during intercourse. This fact raises a question regarding evolutionary accounts: Should they explain the existence of orgasm in general? Or is there some prima facie reason to focus exclusively on the existence of orgasm with intercourse? Every explanation in this chapter attempts to explain the existence only of orgasm with intercourse, and not the existence of female orgasm per se. But what is the justification for this? It would seem to be that only through tying orgasm to intercourse do we get a connection between orgasm and reproductive success. In order for female orgasm to be an *adaptation,* like sickle cell anemia and not like the male nipple, it must be connected with reproductive success, which thereby intuitively links it to reproductive sex, or intercourse. But this assumes that female orgasm *is* an adaptation, a basic assumption that is left completely unsupported. There is no plausible evidence that links orgasm to reproductive success (see Chapter 7). Thus these explanations are all flawed because they assume that orgasm is tied to reproductive success without any supporting evidence. This assumption is manifested in the fact that female orgasm is considered not by itself, but only in relation to intercourse.

Daniel Rancour-Laferriere also addresses the orgasm/intercourse discrepancy, and adds a couple of interesting twists to a pair-bond account. He begins with the assumption that loss of estrus in human females is unique. Given that loss, he writes, "It follows that *something* has to have induced females to practice sexual intercourse, else they would have left significantly" fewer progeny than females with estrus. "That is, there has to have been a preadaptive device which insured intercourse even after estrus was lost" (1983, p. 321; his emphasis).

Technically, a "preadaptation" is one of several things. It is either a trait that arose by chance variation or by its genetic linking with another, selected trait; or it is a feature that was directly selected for one function that can be coopted to serve another function (see Gould and Vrba 1982). In this case, the claim is of the latter sort, where female orgasm existed before the loss of estrus, and enabled loss of estrus to survive as a trait. But since Rancour-Laferriere believes that female orgasm itself is also an adaptation for some other function, he needs an account of its particular selective advantage. Strangely, he omits such an account. Instead, he focuses on how the female orgasm, once fixed in the population, might have been coopted to ensure intercourse after loss of estrus. In comparison to polygynously inclined males, who supposedly only displayed interest in females showing vigorous estrus, when the signals of estrus started to wane in evolution, "more indulgent, paternally inclined males would have succeeded in copulating with females in dampened estrus—provided of course the females themselves had something (such as orgasm) to keep them interested in having sex despite waning estrus" (1983, p. 322). According to Rancour-Laferriere, the female would have evolved to understand that frontal copulation "would be more likely to give her pleasure, including orgasm. She would thus have been motivated to engage in sexual activity regardless of whether she was ovulating or not. Her orgasmic ability had what might be termed *a hedonic function*" (1983, p. 322; his emphasis).

Rancour-Laferriere acknowledges that frontal copulation does not ensure female orgasm, "but even if orgasm was only achieved some of the time the females might have found frontal copulation rewarding" (1983, p. 323). That was either because she enjoyed intercourse for its own sake, in which case the entire hypothesis about female orgasm is superfluous, or because she was, behavioristically speaking, on an intermittent reinforcement schedule (see Chapter 4).

Rancour-Laferriere makes an apparently good point when addressing the orgasm/intercourse discrepancy: "All that the early hominid female (constant or intermittent) orgasm needed, in order to qualify as an ordinary Darwinian adaptation, was to result in greater average reproductive success than the complete absence of female orgasm did" (1983, p. 323). His point is that "the absence of orgasm in a minority of women does not necessarily testify to the non-adaptiveness of female orgasm—any more than impotence or premature ejaculation in a minority of males testifies to the non-adaptiveness of the male orgasm" (1983, p. 324).[7]

This parallel with male orgasm does not really work: most premature ejaculation (that is, ejaculation in the vagina but too soon for the woman's satisfaction) would be just as effective at reproduction as regular orgasm. If we consider impotence instead, the percentage of (especially older) men with impotence is comparable to the percentage of women who never have orgasm. But the cases differ fundamentally, because there is no convincing evidence that female orgasm is related to fertilization.[8] Moreover, with the selection pressure Rancour-Laferriere describes, there is no reason the trait of female orgasm with intercourse would not become universal in the population. Yet it is not, and Rancour-Laferriere has failed to explain how or why.

In addition to his explanation that female orgasm evolved in order to reinforce engaging in intercourse, Rancour-Laferriere gives a second explanation, which he calls the "domestic bliss function." The idea is that female orgasm serves as a signal to females regarding which male to choose to copulate with regularly. He notes that some contemporary studies show a correlation between a woman's satisfaction (happiness) with her marriage and her orgasmic consistency. This happiness, he concludes, helps the female choose the male who is most likely to "stay around and benefit her offspring in the long run" (1983, p. 329). In other words, her own orgasm contains information about whether her marriage is happy, which tells her whether to keep her current mate or not. One problem with using the data on marriage happiness and orgasm is that, notoriously, no one has been able to tell which trait causes which: Does frequent orgasm make the marriage happier? Or does a happy marriage make the woman more orgasmic? But perhaps for Rancour-Laferriere's hypothesis, the causal direction does not matter: the basic rule might be, if the orgasm rate is low, seek another mate. But this will not help with the chief problem with this "domestic bliss" explanation: the unsupported assumption that evolutionarily successful fathering of offspring is associated with sexual performance that causes female orgasm.

John Alcock (1980) also argues that female orgasm serves as a means of female choice of mate. Like Rancour-Laferriere, Alcock suggests that females can use information about whether they had orgasm with intercourse to discriminate among male partners. Females would have chosen to copulate more often with a male with whom copulation leads to orgasm (1980, p. 182). Alcock writes, "This could help a female initiate and maintain a long-term cooperative relationship with a male partner, usually but not necessarily her husband. According to this view, the pleasure produced by an orgasm is one of the many proximate factors that females can *potentially* use to influence their selection of the father of their children. If female orgasm were automatic, a woman could not use its occurrence or absence to help assess the emotional quality of a sexual relationship" (1980, p. 182; his emphasis). Like Rancour-Laferriere's, this explanation has the virtue of acknowledging the orgasm/intercourse discrepancy, but it is only speculation that females choose the fathers of their children on this basis. Note the empirical assumptions made in this hypothesis. In order for it to be true, men would have to show a correlation between their attentiveness as lovers and their responsibility as parents, but there is no evidence of such a correlation (see Mitchell 1992). In addition, there would have to be a correlation between a woman's having orgasm with a partner and her having intercourse with him. Although this may be intuitively appealing, no evidence supporting it has been found. Finally, Alcock assumes that women use the presence or absence of orgasm to evaluate the emotional quality of a sexual relationship, but there is no evidence to support this, either. Finally, this seems to be an explanation of only the 42% of women who sometimes do and sometimes don't have orgasm with intercourse. Under this hypothesis, women without orgasms with intercourse, and women who always have orgasms with intercourse, would be selected against.

## Summary

Claims that the evolution of female orgasm rests on its role in strengthening the pair bond were the mainstay of early evolutionary explanations of the trait. The chief empirical problem with such accounts is the presence of the substantial gap between the occurrence of intercourse and the occurrence of female orgasm (see Table 1). Most accounts simply assume that the function of female orgasm is to play some role in intercourse, a central assumption that is undermined by all of the available data. This fault alone imperils the scientific plausibility of such explanations. A couple of the pair-bond accounts do address this problem, although their approaches are also inadequate; they tend to focus on explaining the evolution of female orgasm during intercourse, as opposed to the evolution of female orgasm per se, which is much more widespread than orgasm during intercourse. An additional problem with all these accounts is that they simply assume, as their basis, that female orgasm is an evolutionary adaptation. But at the time these authors were writing (and arguably now, as well), there was no evidence that linked the trait of female orgasm with the differential reproductive success of females that manifested it. In short, all of these accounts attempt to answer the question, "What is the adaptive function of the trait?" and this begs the question whether or not female orgasm has an adaptive function or not. Problems with this assumption are detailed in Chapters 6 and 7. There are, however, a variety of other early explanations that were offered for the evolution of female orgasm, and I shall address them in the next chapter.

# CHAPTER 4

# Further Evolutionary Accounts of Female Orgasm

In this chapter I review non–pair-bond adaptive explanations and female-centered explanations of the evolution of human female orgasm. The first of the explanations essentially denies the evolutionary existence of the orgasm/intercourse discrepancy. The following three non–pair-bond explanations are distinguished by their lack of evolutionary logic. The female-centered explanations have problems, as well.

## Non–Pair-Bond Explanations

### Melvin Lloyd Allen and William Burton Lemmon's Account

Basing their view on the comparative evidence from nonhuman primates, Allen and Lemmon begin with the assumption that female orgasm is not unique to human beings. Instead, it is one of several discrete sexual responses, each of which may be adaptive in its own right for the primate lineage; the features of female orgasm that they suggest are adaptive include perivaginal muscular contractions, uterine suction, and reflex ovulation. Allen and Lemmon consider each of these adaptive because "any one of these responses could mean the difference between a fertile and an infertile copulation when the male is capable of suboptimal reproductive effort—

e.g., when he has a low sperm count because of recent or frequent ejaculation (a common situation in sexually dimorphic species)" (1981, p. 24).

Although they are aware that many researchers stress that women do not need to have orgasm in order to become pregnant, they also find accounts that rest on female pleasure alone implausible: "does the capacity for female orgasm exist in our species for no other reason than to allow the female to be able to experience highly pleasurable sensations during sexual intercourse? If so, it is intriguing that ecstatic sensations are felt during the contractions of the perivaginal musculature, which seems to effectively grip and massage the penis which may often result in ejaculation in a highly aroused male" (1981, p. 24). (Wendy Faulkner endorses the same explanation; 1980, p. 152.)

This appears to be an argument from incredulity. But there is a new hypothesis being proposed here: "that the orgasmic response in the anthropoid female (Infraorder: *Catarrhini*), or perhaps in mammals generally, has evolved for the purpose of stimulating the orgasmic response in the male. . . . That is, the rhythmically contracting muscles surrounding the vagina serve as the adequate stimulus for ejaculation" (1981, p. 24).[1] Allen and Lemmon offer several bits of evidence for this. They first cite Bartlett (1956), who collected heart-rate and respiration data on coital partners. Allen and Lemmon emphasize that, according to Bartlett's data, "the female orgasm, as signaled by the woman, invariably preceded or accompanied the male orgasm" (1981, p. 24). I would note that Bartlett's data were the results of three trials performed on each of three couples (1956, p. 469). A close look at Bartlett's (notably sparse) data, however, reveals that not all of the female orgasms actually preceded or accompanied male orgasm. Nor do Allen and Lemmon take into account any voluntary control that the males might be exerting with regard to their timing of orgasm (Kinsey et al. 1948, pp. 580–581). Also, the data on the women cannot be considered representative, first, because there are only three women involved, and second, because they all were selected partly because, unlike most women, they always have orgasm with intercourse.

Allen and Lemmon also cite evidence from rhesus macaques, in which females often reach back and clutch the male around the time of his ejaculation. Although at the time Allen and Lemmon were writing, the identification of the "clutching reaction" as an indication of female orgasm was still a live hypothesis, more recent evidence from electronic studies of vaginal contractions in monkeys has shown that the clutching reaction should not be identified with female orgasm (Slob et al. 1986, p. 894).

Allen and Lemmon's final piece of evidence supporting the claim that female orgasm evolved to stimulate male orgasm is that when one is collecting semen for artificial insemination, rhythmic contractions are necessary in order for male ejaculation to occur in both the boar and the dog. I think that the fact that these rhythmic vaginal contractions occur in nonhuman primate females without the occurrence of female orgasm undermines Allen and Lemmon's assumption that these contractions must be a consequence of orgasm (Slob and van der Werff ten Bosch 1991, p. 141).

One strength of their hypothesis is that Allen and Lemmon do address the orgasm/intercourse discrepancy. Why, they ask, "do so many women fail to reach orgasm at all during coitus, or reach it only after a considerable duration, and the male still reaches orgasm without being stimulated by the female's rhythmic vaginal contractions?" (1981, p. 25). They answer that female orgasm, as an adaptive response, has not been selected to occur during every possible copulation, but, rather, it functions "as a means of *female choice* after penetration has been achieved" (1981, p. 25; their emphasis). The idea is that female orgasm is supposed to increase the chance that the male will have an orgasm; if the woman can then refrain from or encourage an orgasm, she may prevent fertilization by undesirable males, or encourage fertilization by desirable males, respectively.

Allen and Lemmon admit, however, that there seems to be something wrong with a simple female-choice hypothesis: that there are "instances when vaginal contractions are not achieved by the female, even when sexual intercourse has been reported to be desirable" (1981, p. 25). (I shall consider related female-choice hypotheses in Chapter 7).

The rest of their article is devoted to showing that, in the natural past evolutionary state, female hominids *did,* in fact, have frequent and reliable orgasm with intercourse, unlike today's females. This is a unique approach to the orgasm/intercourse discrepancy: show that the discrepancy itself is only a recent phenomenon, and that our ancestral females were much more orgasmic than today's human females.

Allen and Lemmon offer several lines of argument for their view about what they call "the incomplete expression of female sexual response." First, they claim that in the evolutionary past males did not have such a low threshold of stimulation to ejaculation, and were thus regularly able to sustain intercourse over longer periods of time; this increased length of intercourse is taken by Allen and Lemmon to have led to increased rates of coital orgasm for the female (but see the discussion of limitations to this argument later in the chapter). They claim that the low threshold to ejaculation is a result of much less frequent copulation now than in the evolutionary past; they use chimpanzees as a comparison species, and speculate that there is a threshold of ejaculation that is specific to human beings, and that it is much higher than today's standard. This speculated ejaculatory threshold means that males would have responded with ejaculation only to "appropriate stimulation"—that from vaginal contractions—and not simply from the "inappropriate sources" of friction from the vaginal walls (1981, p. 25).

According to Allen and Lemmon, men today cannot help having a lower ejaculatory threshold, because they do not have the extra opportunities for copulation that would be necessary to achieve the "species-typical" level. But, according to them, women have their own psyches to blame for their low rate of coital orgasm relative to that of our prehistoric ancestors. Allen and Lemmon write, "It has been well documented that many women cannot reach orgasm as quickly as can the average man because of psychological inhibition" (1981, p. 25). In support of this claim they cite Freud (1905), several Freudians, and others. Astonishingly, they seem to be completely oblivious of the work of both Kinsey and colleagues and Masters and Johnson, who argued that the difference in orgasm frequency between men and women was due to the lack of stimulation received by women during intercourse, and not to psychological inhibition. There have been numerous studies attempting to isolate psychological differences between women who have regular coital orgasm and those who do not; the results produced have been equivocal, and have shown minimal psychological differences.[2] Still, what is most strange about Allen and Lemmon's appeal to Freud is that they do not even bother addressing the view that a mechanical lack of stimulation is primarily responsible for the orgasm/intercourse discrepancy, certainly the overwhelmingly dominant view in sexology at the time they were writing.

Allen and Lemmon develop a second account of the contemporary shortfall of coital orgasm by women, which is explained by focusing on their muscle condition. They write that the lack of control and development of the pubococcygeus muscles is due to restrictive societies, in which women do not learn to operate and strengthen their vaginal muscles (1981, p. 26). This seems to imply that past societies were less restrictive, and more oriented toward female sexual satisfaction. Not only is there no evidence for this, but the historical and cross-cultural information available brings this into doubt, because standard approaches to sex in a variety of societies neglect female orgasm entirely (Davenport 1977).

Allen and Lemmon draw attention to the important fact that women masturbate to orgasm in just a few minutes, noting that masturbation focuses on clitoral stimulation, which is a much more speedy and consistent way of producing orgasm than coitus (1981, p. 26). Allen and Lemmon then leap to the conclusion that "no more than about five minutes should be required to induce orgasm in a sociobiologically compatible pair during heterosexual coitus, unless it is purposefully delayed to prolong sexual arousal" (1981, p. 26).

It is unclear what Allen and Lemmon could mean here. They seem to be manufacturing their own artificial standard for the proper timing and frequency of female orgasm and then to be holding the women (and their societies) who fail to meet their standard responsible. Apparently, the documented brevity of stimulation required for orgasm upon masturbation is being used by them to shore up their claim that today's women are psychologically flawed, socially repressed, and too inadequately muscled to achieve the "species-typical" state that Allen and Lemmon have made up. The burden of proof is on Allen and Lemmon to show that there was such a different "species-typical" state in the evolutionary past, and they have not even tried to shoulder it.

Allen and Lemmon, in the end, offer one more evolutionary reason for female orgasm, but they do not discuss it at length. They propose that it would also be adaptive if the sperm were aided in their travel up the cervical canal and into the uterus, and that uterine suction, induced by female orgasm, may produce this effect (they cite Fox et al. 1970 and Singer 1973). I shall discuss this uterine-suction hypothesis at length in Chapter 7.

All told, Allen and Lemmon's account relies heavily on the notion of "species-typical" responses in ancestral men and women that are quite different from what has been documented. They argue that female orgasm was an adaptation to assist the male to ejaculate. This requires that the male increased his reproductive success by receiving help ejaculating, that the females regularly had orgasms with intercourse, and that they did so prior to male ejaculation. Only the last of these assumptions perhaps holds true today in some cultures, but Allen and Lemmon claim that all of the assumptions were true in the evolutionary past, while providing no evidence to support them. Perhaps most striking is their complete neglect of the sexologists' claim that intercourse does not provide enough mechanical stimulation in females to produce consistent orgasm. Allen and Lemmon are apparently unworried by the conflict between their own representations of female sexuality and what the sexologists have actually documented.

### William Bernds and David Barash's Account

Bernds and Barash's is probably the most illogical of all the accounts of the evolution of female orgasm. Their account says that there is a selective advantage to women who spontaneously abort their fetuses upon experiencing orgasm. Such abortions are themselves taken to be adaptive because environments in which such abortions optimize the females' allocation of resources were sufficiently common in the evolutionary history of human beings. The prime example of such an environment is one in which a hostile male is prone to kill an offspring that is not his.

Bernds and Barash assemble their evidence for this view from a number of sources. Their story is based on the claim that there is circumstantial evidence that women's orgasm may induce abortion (to be addressed below). They then assume that the existence of women's orgasm calls for an adaptive explanation (1979, p. 500). They note that female rats reabsorb their litters if a strange male is present: upon exposure to the new male, pregnant rats will absorb the tissue of the fetuses, thereby becoming not pregnant. This is taken to be adaptive because it saves the females having to bear the litters, only to have them killed by the new male. By reabsorbing their litters, they can more immediately become receptive to the new male, and not waste their resources on a doomed litter.

Bernds and Barash then write, "Throughout human history it has not been unusual to take women as spoils of war (Mead 1950). If the subsequent offspring were subject to infanticide, those women who aborted as a consequence of orgasm would be at a selective advantage" (1979, p. 500).

First, let us examine the "circumstantial evidence" they offer for orgasm's inducing abortion. They cite Javert (1957), who noted that Kinsey's data on the percentage of coital orgasms by age correlated fairly well with the age distribution of spontaneous abortion (1979, p. 500): women's rate of orgasm increases with age, as does the rate of spontaneous abortion. But correlation is not cause, and having two traits vary together does not necessarily mean that one caused the other. Moreover, one can easily cite independent causes, such as experience and aging, respectively, that would account for these data.

Bernds and Barash also appeal to Masters and Johnson's claim that *if* it is true that orgasm from coitus in pregnant females tends to cause miscarriage, then so would orgasm from masturbation. But their appeal is highly misleading, since Masters and Johnson explicitly *deny* that orgasm from coitus tends to cause miscarriage (1966, p. 165).

Bernds and Barash attempt to use the data showing that a larger percentage of women nearing the end of their pregnancy "experienced a significant reduction in eroticism and frequency of sexual performance," to infer that this behavior confers a selective advantage (1979, p. 500). But one does not need a fertile imagination to understand a statistical falling off of sexual interaction as delivery approaches; just because a phenomenon exists does not mean that it is an evolutionary adaptation. Bernds and Barash also cite Masters and Johnson's claim that "it is probably true that contractions of orgasm *at or near term* can send a woman into labor" (Masters and Johnson 1966, p. 166, my emphasis; Bernds and Barash 1979, p. 500). Bernds and Barash use this to conclude that "the data imply a possible selective disadvantage to women who are orgasmic in their third trimester of pregnancy" (1979, p. 500). But what would be the selective disadvantage of delivering "at or near term"? This phenomenon hardly qualifies as "evidence that orgasm may induce abortion," unless they believe that birth is the same as abortion.

Their adaptive story suffers from illogic as well. Not only do Bernds and Barash adopt the implausible assumption that women have enough orgasms while being raped to make this evolutionary strategy effective; it seems that, in order to make Bernds and Barash's account work, women must have orgasms *only* (or mostly) while being raped by their conquerors. Otherwise, they would have orgasms (and therefore miscarriages) in the context of established relationships, which would certainly *lower* their fitness more than spontaneous abortion upon being conquered would *raise* it (see Kitcher 1985, p. 239).

In summary then, there is no good evidence for Bernds and Barash's fundamental assumption, that orgasm induces abortion in women. Their account also contains the completely undefended assumption that women would routinely have orgasm during rape. Finally, following their own logic, one can infer that orgasm outside of rape would be selected against. What is perhaps most striking about this account is that it was published in a landmark and widely cited edited volume (Chagnon and Irons 1979).

## Richard Alexander and Katharine Noonan's Account

Richard Alexander and Katherine Noonan argue against the view (promoted by Campbell, Morris, Crook, and others) that females pair with males by supplying constant sex. The problem with the pair-bond maintenance theories, they claim, is that they do not explain why males would want frequent sex with no hope of fertilization (1979, p. 443). Instead, they offer two explanations for female orgasm. On one explanation, the female orgasm has the evolutionary function of communicating a possible abortion to the male. On the other explanation, the female orgasm signals to the male that the female is satisfied.

Alexander and Noonan do recognize the existence of female orgasms in other species, but they also claim that the intensity and presence of outward signs of orgasm may be unique to human beings. They examine the conditions under which it would be most adaptive to show the external signs of orgasm seen in human beings (1979, p. 450). Note that they are trying to explain what they take to be a uniquely human trait, the existence of "external signs of orgasm" rather than orgasm itself. As we shall see in Chapter 5, nonhuman primate females also show external signs of orgasm, but let us see what their hypothesis for the human case is.

First, they suggest that external signs of orgasm would be adaptive because "orgasms may sometimes increase the likelihood of abortion, thus decreasing the likelihood of paternity mistakes in some circumstances" (1979, p. 450; see the objections raised above). This implies that the smart male should try to give his mate(s) orgasms only when there is some doubt that he is the father, and *not* at all other times, when he might risk the abortion of his own offspring. Thus female orgasm should be relatively rare. One may ask, why would this encourage the "display" of female orgasm? Presumably, Alexander and Noonan's answer would be: in order to demonstrate to the male that he has possibly induced an abortion. Given the health risks of abortion to the females, it would seem that those women who *pretended* to have orgasms, but actually did not, would be better off.

Second, Alexander and Noonan also suggest that the external display of orgasm would be selected in females because it "may communicate the female's sexual satisfaction to the male" (1979, p. 450). More specifically, they claim that the resemblance of female orgasm to male orgasm suggests to the male that the female has experienced a satisfaction similar to his (1979, p. 451). The reason such communication is supposed to be adaptive for the female is that it is supposed to protect her from retaliation from a male suspicious of the woman's fidelity. Thus Alexander and Noonan suggest that if the female orgasm resembles male orgasm, the resemblance "may be a communicative device which tends to raise her male's confidence that the female is disinclined to seek sexual satisfaction with other males" (1979, p. 451). But this argument works on the strength of several questionable assumptions. First, it assumes that the males can tell whether the female has had an orgasm, a questionable assumption, given the high incidence of successful faking by women (see, for example, Hamilton 1929; Laumann et al. 1994; Thornhill et al. 1995). It also assumes that the male is interested in the female's sexual satisfaction, which contradicts some of the cross-cultural evidence (Davenport 1977; see the discussion in Chapter 5). Also, it must be true that women's sexuality is like men's; but almost half of the women in Hite's study (which had already been published when Alexander and Noonan published their theory) desired more than one orgasm on many occasions, although it is widely true that men are satisfied by a single orgasm. Furthermore, there is some evidence that the more orgasms a woman has, the more she desires sexual activity (Masters and Johnson 1970).[3] If Japanese macaques are a relevant animal model—an assumption that Alexander and Noonan should accept, since the species fits their specifications of which animal models are relevant (1979, p. 450), then their argument does not work at all: Wolfe found that those adult females who mounted males (thereby stimulating themselves) and showed more excitement "had a statistically significant greater number of male partners . . . than those females . . . who did not" (1979, p. 528). Hence, the male might have little reason for his confidence that the stimulated female will not go looking for fresh partners. As in Pugh's account, it seems that the male's purpose would be better served if the female did not have orgasms at all, or at least if she did not have them with intercourse, so she would be less likely to be impregnated by another male.

Alexander and Noonan continue by predicting that if their interpretation is correct, the outward signs of female orgasms should mimic male orgasms, and "frequently involve deception, with females pretending to have orgasms when they do not" (1979, p. 451). Thus Alexander and Noonan's arguments can just as well be viewed as adaptive stories for *faking* orgasm, especially given the supposed possible risks of abortion that they mention.

At any rate, according to Alexander and Noonan, the highest rates of orgasm should appear in women who are in "deeply satisfying or long-term interactions with males committed to the female and her offspring" and "with dominant males or males with obviously superior ability to deliver parental benefits" (1979, p. 451). In addition, female orgasm would be least likely in "brief or casual encounters." These particular predictions follow from their overall adaptive story: "female orgasms would perhaps be more likely in females trying to obtain or keep a 'good' male, identified as a male with much parental investment to offer. Thus, we believe that this feature of human female sexuality [outward signs of orgasm] too may be linked to male parental care and the threat of desertion" (1979, p. 451).

There are several problems here. First, Alexander and Noonan offer no evidence to support their prediction that orgasms should occur more frequently with a "dominant male," though they claim vaguely that their predictions are "compatible" with the evidence about human female orgasm. Elsewhere Alexander does offer documentation of the claim that females prefer high-ranking males, mentioning one study in which high-ranking human families reproduce more than low-ranking ones, but none of this addresses the claim at hand, that females are more likely to have orgasms with dominant males (Alexander et al. 1979).

Furthermore, Alexander and Noonan conclude that orgasm is supposed to help "obtain or keep" a good male. But Kinsey and Gebhard separately showed that the female's orgasmic frequency increases over a 20-year period (documented by Gebhard 1966 and also Kinsey et al. 1953, and cited by Alexander and Noonan). This initially low orgasmic capacity certainly cannot help (and could conceivably hurt) in *obtaining* a good mate, and it is difficult to see that it would help the female to *keep* him either, given that the increase in response is very gradual, and most likely to show after she is past any serious dependency. In other words, if the female is already in a long-term committed relationship, in which she is likely to have already raised her offspring past their (most demanding) infancy, it is not clear why it would help her fitness significantly to demonstrate her devotion to her mate through an increased frequency of (the appearance of) orgasm.

In summary, Alexander and Noonan fail to say why their account is not an adaptive story for faking, rather than an adaptive story for an intense and visible female orgasm. Even if we read it as an account of real female orgasm, furthermore, they take no notice of the orgasm/intercourse discrepancy, which would seem to be damaging to their case, since they are concerned only with explaining the phenomenon of orgasm with intercourse.

## The Intermittent Reinforcement Account

Milton Diamond and Sarah Blaffer Hrdy offer an intriguing explanation of the evolution of female orgasm based on experimental psychology. Diamond begins by recognizing the orgasm/intercourse discrepancy. His view is that the wide variability in female orgasmic response during intercourse is *itself* an adaptation. He starts by observing that, even among healthy and well-fed Americans, about one in eight couples has infertility problems, many of which are due to a deficiency in sperm quality, quantity, or transport. Thus a higher rate of intercourse would be an important feature of ensuring successful fertilization: "[Evolving] human females could, based on operant learning theory, be driven harder by irregular reward (orgasm) to seek sexual satisfaction. Partial and irregular re-inforcement is known to produce more persistent behavior than 100 percent reward. Thus frustration might be satisfied by seeking or accepting additional coitus or different partners ensuring a more regular or varied supply of sperm to help ensure conception and the survival of their gene pool" (1980, p. 184).

This is an echo of an argument given by Sarah Hrdy (1979, p. 312). The principle of intermittent reinforcement states that be-haviors that receive unpredictable rewards are very stable and dif-ficult to extinguish. Thus even the highly variable female orgasm could serve as an effective reward system for engaging in sexual in-tercourse.

This argument, however, does not work. Intermittent reinforce-ment schedules are only effective if an initial schedule of regular re-inforcement was already in place, wherein the subject gets a reward every time it shows the behavior. That is, you cannot train a rat to hit the lever 20 times to get a food pellet to appear, unless you first train the rat to hit the lever by rewarding it each time it does any-thing like lever pressing (see Domjan 1998, pp. 128–129). Among women, that would mean that women would not repeat the behav-ior of intercourse an increased number of times without orgasm, unless they first were under a schedule in which each act of in-tercourse were accompanied by orgasm (a claim contradicted by Kinsey et al. 1953). Now, there are certainly other motivations be-sides orgasm to induce a woman to have intercourse, but the hy-pothesis in question is proposing that orgasm is the reward for in-tercourse. Thus the psychological dynamics of this explanation do not really work.

In addition to this central problem, I would like to draw atten-tion to an ambiguity here. When female orgasm is called highly variable, this could mean either that individual females vary widely in whether they experience orgasm on a particular occasion of intercourse—sometimes they do and sometimes they don't—or it could mean that different females vary in whether they tend to experience orgasm with intercourse or not. As we know from the sexology literature, both of these types of variation are widespread. But in order for the intermittent reinforcement theory to work, it is only the first meaning that can be in play. In other words, this argument must focus on those females that sometimes do and sometimes don't have orgasm with intercourse. Such women make up only from 36% to 44% of the population (Kinsey et al. 1953, p. 375; see Chapter 2). Thus it seems that although a sizable percentage of women are included in this hypothesis, and the hypothesized mechanism of intermittent reinforcement can apply to these women, it does not apply to the majority of women. It cannot be used to explain orgasm in general.

Also, what are we to make of Diamond's statistics about infertility? The frequency of abortion and infanticide among women worldwide militates against any scenario involving a need to increase the fertilization rate, because, under circumstances of widespread abortion and infanticide, increased fertilization results in wastage and endangerment of women's health. Diamond gives no citations for his infertility statistics, and I suggest that late twentieth-century American fertilization rates may not in any case be representative of evolutionarily normal rates. Some researchers have documented a sharp reduction in sperm count over the latter part of the twentieth century, believed by some to be caused by the persistence of hormone-like pesticides in the environment (Duty et al. 2003; Swan 2003).

In the end, although Diamond's explanation does have the advantage of dealing with the orgasm/intercourse discrepancy straight on, the intermittent reinforcement account suffers from problems with its own logic, as well as not applying to even half of women.

## Female-Centered Explanations

Two of the early explanations for female orgasm revolve around the reproductive advantage to the female alone, rather than to the female-male combination. The first of these accounts is based on the health advantages of orgasm; the other is based on the concept of sexual selection, but not in one of its two usual forms of female choice and male-male competition. Instead, this second account focuses on another dynamic in selection, competition for breeding success between females.

### Mary Jane Sherfey's Account

Sherfey proposes a health advantage for the female as the selection pressure driving evolution of the human female orgasm. She argues that female orgasms have been specially adapted to serve the therapeutic purpose of relieving vasocongestion in the pelvis. According to Sherfey, selection pressure favored the development of longer orgasms in women than in men (documented by Masters and Johnson) because longer orgasms ensure "the expulsion of the greatest amount of pelvic vascular congestion" (1973, p. 80). Sherfey claims that this removal of massive pelvic congestion (caused by sexual excitement) is important, since "chronic pelvic congestion furthered by inadequate or absent orgasmic release fosters a pelvic condition conducive to many disorders interfering with impregnation, pregnancies, and general health" (1973, p. 80; see Taylor 1962, p. 472). She draws the conclusion that orgasm in women is adapted to "remove the largest amount of venous congestion in the most effective manner" (1973, p. 80).

Sherfey also emphasizes the capacity of women to have and want a large number of orgasms in a relatively brief period of time. She notes that "with full venous engorgement and edema" the areas emptied by orgasm quickly refill almost immediately thereafter; continued stimulation soon stretches the muscles again to the point of reflex contraction, and another orgasm occurs. "Hence," she argues, "the necessity and capacity for three to six or many more orgasms in a short time are explained, as well as why no orgasm (or a single small one) leaves a woman who has been adequately aroused feeling uncomfortable, and why it takes so long for sexual tension to subside when no orgasm occurs" (1973, p. 105). This, combined with the fact that increased sexual activity over time yields increased vascularization (presence of blood vessels) of the pelvic area, which in turn corresponds to increased ability to attain high levels of arousal, leads Sherfey to conclude that "the more orgasms a woman has, the more she can have. To all intents and purposes, *the human female is sexually insatiable in the presence of the highest degree of sexual satiation*" (1973, p. 112; her emphasis).

The result of this sexual interest is supposed to be that it creates the inclination of a woman to seek partners and to persist in copulation until sexually satisfied, thus producing the healthful reduction in pelvic congestion. But on the description just given, it is unclear that this decongestion would ever occur in a way likely to have long-term benefits; Sherfey's description of female insatiability seems to imply that it would not. Thus these two threads of Sherfey's account are in tension. Either the woman can acquire enough orgasms to bring about the decongestion of the pelvic region, or she is sexually insatiable once aroused and incapable of such decongestion; they are not compatible. The charitable reading of Sherfey is that she is committed to the decongestion adaptive account, and simply got carried away when describing female sexual drive.

In any event, Sherfey, in the process of performing the invaluable service of bringing Masters and Johnson's bad news about Freudian "vaginal orgasm" to the psychiatric community, seems to have been overly impressed by the indirect method of attaining orgasms during intercourse, described by Masters and Johnson. Sherfey argues, in great anatomical detail, for the *perfect adaptation* of the female external genitals to stimulation by the thrusting of intercourse. For example, she claims: "the fantastic variations [which, she predicts, will not be found in other primates] in the glans-preputial-labial

anatomy [in which the clitoral glans or head is rubbed by the pre-
puce (hood) through pulling on the labia], all with the purpose of
maintaining or improving the frictional action of the glans against
the inner layer of the preputial mucous membrane, are impres-
sive evidence of the extent to which evolution has gone to *perfect*
this mechanism. Its importance to the successful performance of
the sexual act seems undeniable" (1973, p. 84; my emphasis). Of
course, as with Masters and Johnson, the fact that the majority
of women do *not* usually have orgasms through this particular
mechanism of intercourse substantially undermines Sherfey's claim
to the evolutionary "perfection" of the mechanism. Even among
the women who do regularly experience orgasm with intercourse,
many of them do so through additional clitoral stimulation by the
hand, and not through the action of the hood rubbing across the clitoris through penile insertion.

Sherfey's adaptationist assumption colors her whole discussion.
After asserting that the selective advantage of male orgasm is obvious, she claims that since the female carried the "bigger burden for
the perpetuation of the species, in reproduction and care of the
young, *there is no logic* in the idea that selection pressure selecting for the reproductive advantage of the orgasmic capacity of the
male, should find this same capacity disadvantageous and unrewarding for the female" (1973, p. 113; my emphasis). But here
Sherfey seems to misunderstand the evolutionary choices available;
claiming that female orgasm is not itself an adaptation does not
presume that it is *dis*advantageous, nor does it imply that it is
behaviorally unrewarding. Female orgasm may be both advantageous to the female and very rewarding: this still does not mean it is
*evolutionarily* advantageous—was itself directly selected (see Chapters 5 and 6). It is also clear that although Sherfey is pursuing an interesting line of thought by examining the health consequences of
female orgasm, she does not have a consistent position on whether
orgasm(s) enables the healthful draining of congested tissues after
sexual excitement. Moreover, while Sherfey claims that orgasm is

adaptive in females because it motivates the female to seek additional partners, she does not explain what the evolutionary advantage of multiple partners would be, exactly (see the discussion of Hrdy, below). Finally, her position on the "perfection" of the indirect stimulation of the clitoris by its hood leading to orgasm is indefensible. It seems likely that Sherfey, like Morris, was misled by the description given by Masters and Johnson of their subject women, who easily had orgasm with unassisted intercourse; these women represent only a minority of women, and their responses to intercourse cannot be used as a standard in evolutionary discussions.

### Sarah Blaffer Hrdy's Account

Hrdy takes up where Sherfey left off. Citing Sherfey's theory that genital pleasure in females is adaptive, in that it creates the inclination to seek partners and persist in copulation until the female is satisfied, Hrdy notes that Sherfey has not actually given us a full evolutionary account. "Adaptive for what?" asks Hrdy; and she proposes a unique and original answer to this question.

Hrdy highlights the fact that sexual selection can occur in contexts other than male-male competition or female choice. She emphasizes that females take many opportunities to increase their chances of reproductive success, and that a mother's social circumstances, for example, the availability of food, helpers, and protection, have great consequences for her offspring's survival. Thus any condition, including the female's sexuality, that affects those social circumstances must be understood as evolutionarily significant. As Hrdy writes, "Only a failure to think seriously about females and to consider the evidence would allow someone to conclude that natural selection operates more powerfully on male sexuality than on female sexuality, or to believe that the female's reproductive character could be 'invisible' to natural selection" (1981, p. 173; cf. Symons 1979, pp. 89, 91, 192).

According to Hrdy, the social factors influencing a female's reproductive success can depend on a number of factors involving male behavior, including the nearby males' tolerance for infants and male willingness either to assist an infant or to leave it alone. In this context, Hrdy proposes, a female's active and promiscuous sexuality becomes selectively important. If females are promiscuous, a male cannot be sure about the paternity of offspring, and is therefore less likely to harm any infant that belongs to a female with whom he mated, because the infant might be one of his own. Thus a female who consorted with 11 males, say, would experience greater protection (and, possibly, care) of her infant than a female that mated with only 3 males. On this scenario, according to Hrdy, the female's enjoyment of sex would serve her interests at the expense of male interests (1981, p. 174). As Hrdy puts it, "Female primates influence males by consorting with them, thereby manipulating the information available to males about possible paternity" (1981, p. 174).

An important feature of Hrdy's hypothesis is that she also takes account of the orgasm/intercourse discrepancy and she portrays it as adaptive.[4] On her account, females obtain their reproductive advantage over other females by being promiscuous. So there must be some mechanism motivating the females to be promiscuous. According to Hrdy, this mechanism is the orgasm/intercourse discrepancy. After noting that only about one quarter of women regularly have orgasm from intercourse alone (1981, p. 166), Hrdy writes: "the physiology of the clitoris, which does not typically generate orgasm after a single copulation, ceases to be mysterious if we put aside the idea that women's sexuality evolved in order to 'serve' her mate, and examine instead the possibility that it evolved in order to increase the reproductive success of primate mothers through enhanced survival of their offspring" (1981, p. 176). She proposes that the sexual excitement generated by the clitoris motivates females to seek *further* copulations in order to attain an orgasm: "it might well be adaptive to possess motivational underpinnings

which would induce females to keep soliciting male partners"
(1988, p. 123). The most significant aspect of this explanation, for
my purposes, is that Hrdy claims that the stimulation of mating
with multiple partners over a brief period of time is *cumulative*
(1996, p. 852). Thus the female will keep soliciting male partners
until she has achieved the level of stimulation necessary for orgasm
(1988, p. 123; 1997, p. 21). Hrdy concludes, "Such a variable reward system as the female orgasm is not as maladaptive as it seems
when we confine our imagination exclusively to monogamous contexts" (1988, p. 123).

Hrdy is fully aware that the breeding system necessary for such
accumulated orgasms from multiple matings is not like any current human breeding system. She is thus not arguing that promiscuity is *currently* adaptive for women. Rather, she is suggesting that
"the female orgasm is *no longer adaptive,* but may be a vestige of a
response that was adaptive in a prehominid condition when our ancestors lived in breeding systems comparable to those of the monkeys and apes" she describes (1988, p. 131; my emphasis). This an-
cient adaptation could, she acknowledges, be subject to secondary
selection incorporating it into more monogamous human breeding
systems. Still, she insists, this would have occurred only long after
the existence of female orgasm itself evolved. Moreover, there is no
good way to tell whether there has been such secondary selection,
because so little work has been done on the comparative anatomy
of female genitalia (1996, p. 852). In sum, then, even if it turned out
that female orgasms are a "legacy" from our primate antecedents,
the trait could either (1) continue to be adaptive in some contexts;
(2) no longer be adaptive; or (3) have evolved in our new context
and by now "have been secondarily enlisted to function in a new
capacity" (1996, p. 852).

One of the most fundamental problems with Hrdy's account is
that it is in some tension with the sexology literature on female sexual response. In particular, it assumes that longer periods of stimu-

lation from intercourse—from multiple males—lead to higher incidence of orgasm.

Only two studies report any correlation between female orgasm
and length of intromission (time that the penis is in the vagina).
Gebhard found that the rate of orgasm with intercourse rose
steadily from 28% of women who have orgasm with 1 minute
or less of intromission to a maximum of 66% who have orgasm
(sometimes) with an intromission time of 15 minutes or longer
(1970, p. 30). Thus intromission time was significant in only 38%
of women in achieving orgasm in this study. Schnabl reports that
25% of women in his study have orgasm with 2 minutes or less of
intromission, and that rises to about 60% after 10 minutes of intromission, which Schabl found to be the mean time for men (1980,
p. 157). He notes that after 10 minutes increased time of intromission made no difference at all. Only 35% of women were affected
by intromission time in Schnabl's study. Both Gebhard and Schnabl
found that the duration and type of foreplay were greater influences
on women's orgasm rate than time of intromission.

Terman found, in contrast, that "the usual duration of intercourse . . . shows little relationship with the orgasmic adequacy of
wives" (1951, p. 165). In a more recent study, no correlation was
found between orgasm rate and time of intromission (Huey et al.
1981, p. 113). Hite's findings were similar (1976, p. 297). Thus the
sex research findings about whether the duration of intercourse
makes a difference to orgasm rate are mixed. Even the two researchers who found an effect showed no increase in the orgasm
rate of women past some point between 10 and 15 minutes. Beyond
a certain point, more is not better. More important, even among the
researchers who found a timing effect, there was an upper limit to
the percentage of women (about 65%, maximum) who could have
orgasm with intercourse at all. Even in the two studies that showed
an effect of intromission time, this effect occurred in only one third
of women. If there is strong directional selection pressure, as Hrdy

supposes, the fact that two thirds of women do not perform as Hrdy predicts undermines her account.

Perhaps, Hrdy might reply, women today do not respond to increased length of intromission time, but this does not mean that they always failed to respond. That is true, and it would help Hrdy's hypothesis if she argued for a past response different from today's, but she does not. Indeed, she would need to show that accumulated time made a difference in the past, but that that response changed due to some change in the selective regime.

The most serious problem with her account, though, is that only females who required increased intercourse time—but still had orgasm then—would be included under Hrdy's account. This leaves the majority of women, who either rarely, never, or always have orgasm with intercourse, out of the account entirely.

Hrdy's scenario does, though, seem like a very good argument for the selective advantage of sexual excitability in some female primates. Sexual excitability, and the role of the clitoris in that excitability, would have been selected through the advantages to offspring that she proposes. It would have led the females to seek sex partners and to confuse paternity. Given this account, however, female orgasm itself would seem unnecessary to the story, especially since it is an unlikely result of prolonging intercourse.

It seems that Hrdy might have been led to her mistaken position on the cumulative effects of intercourse by her reading of the studies on nonhuman primate females. She discusses Frances Burton's experiments on female orgasm in the rhesus macaque, in which Burton induced orgasm through manual stimulation of the clitoris and artificial vaginal penetration. Hrdy notes that "levels of stimulation comparable to those which induced orgasm in the laboratory would occur in the wild only if there were multiple copulatory bouts and sexual stimulation was cumulative from bout to bout" (1981, p. 169). As Hrdy observes, female rhesus macaques do have repeated intercourse with a variety of partners in the wild; she concludes, "there exists a distinct possibility that stimulation sufficient for orgasm occurs in the wild" during periods of intercourse (1981, p. 170). But this is unsupported by the available evidence. Although it is clear that stumptail macaque females occasionally show the distinct outward signs of orgasm during copulation, there is no evidence to support the claim that rhesus macaques also do so. The previously held belief that macaque female orgasm may be signaled by the "clutching reaction" has been debunked (Slob et al. 1986, p. 894). Burton herself concluded that her own experiments indicated that female orgasm *did not* occur in the wild. This finding was based on the duration of a copulatory bout—a series of mounts leading to ejaculation—in the wild. Burton argues that the prolonged clitoral and vaginal stimulation provided in the experiment are not replicated in normal copulation.

But Hrdy has a point: Burton did not consider the very common scenario in which a female at peak estrus will copulate with one male after another. If sexual excitement is cumulative, then perhaps the female would acquire enough stimulation to attain an orgasm. There are two problems with this, though: first, Burton's experiment included 5 minutes of direct clitoral stimulation, and it is not clear that the rhesus females could replicate this without themselves mounting either males or other females, as the stumptails do; and second, once again, no observation of female orgasm during copulation has ever been recorded among the rhesus macaques. So while Hrdy seems to think that the cumulative stimulation theory is supported by the rhesus female orgasm, that is not actually true. We have more reason to believe that rhesus females respond like human females to increased intromission time—it leads to an increase and maintenance of sexual excitement, but not to orgasm.

Though Hrdy claims that "orgasmic 'reward' systems dependent on prolonged stimulation [are] now documented among rhesus and stump-tailed macaques," citing Burton 1971, Goldfoot et al. 1980,

Slob et al. 1986, and Slob and van der Werff ten Bosch 1991, none of these studies actually supports her claim (1997, p. 21; see Chapter 5 for further discussion).

Hrdy's use of the nonhuman primate data sheds some light on her overall approach. What we have from rhesus and other nonhuman primate studies is information that female sexual arousal may lead to continued interest in copulating with a variety of males. Evolutionarily, then, the selection pressure must be understood as impinging on the clitoris and the innervation of the female genitals to increase and maintain sexual excitement. But there is as yet no call for discussing orgasm in this context. Nevertheless, Hrdy links the function of the clitoris to the evocation of orgasm: "Are we to assume, then, that [the clitoris] is irrelevant? . . . It would be safer to suspect that, like most organs . . . it serves a purpose, or once did. But the purpose . . . appears to be transmitting the pleasurable sexual stimulations that sometimes culminate in orgasm. And that brings us full circle: to rationalize the existence of a clitoris in evolutionary terms, we must show that female orgasms confer some reproductive advantage on the creatures experiencing them" (1981, p. 167).

Here we see a confusion on Hrdy's part. No one has argued that the clitoris does not serve the evolutionary function of creating and maintaining a level of sexual stimulation or excitement necessary for intercourse. The existence (or not) of orgasm is a separate matter; the evolutionary function of the clitoris can be treated completely separately through focusing on sexual excitement. This approach is supported by the nonhuman and human data regarding the role of the clitoris in sexual excitement. Hrdy's linking of the evolution of orgasm and the existence and responsiveness of the clitoris is thus inappropriate without further explanation.

A further difficulty with Hrdy's explanation is her assumption that females may influence the survival of their offspring by manipulating the information available to the males about paternity. This assumption has been confirmed among birds, but not among the primates for whom her explanation was proposed (1997, p. 18). And, as Hrdy writes, "The hypothesis that a female might inhibit males from subsequently attacking her infant, or else elicit extra protection or support for her offspring by casting wide the net of possible paternity, predicts that past consortships do indeed affect male behavior towards the offspring of former consorts" (1997, p. 18); but Hrdy admits very frankly that the only evidence supporting this claim is the same evidence that inspired her hypothesis in the first place. More specifically, this is the evidence from savanna baboons, barbary macaques, and human beings that males look out for the well-being of infants (not necessarily their own), and that infanticide by adult males is most likely to occur when a male enters the group from outside, thus guaranteeing that he is not related to the infant he kills. There is also evidence from langurs that males newly entering a group kill offspring (1997, p. 18). The problem for Hrdy is that the view that killing of unrelated infants by males in the group could have persisted into the hominid line has been criticized by Strassman. With the advent of cooperative hunting, Hrdy argues, there was strong selective pressure for group cohesiveness; males who had to cooperate with one another were very unlikely to be killing one another's infants (1981, p. 32; this view is reinforced by Crook's findings; 1972, p. 270). Moreover, evidence that is problematic for Hrdy's theory comes from the fact that most cases of human infanticide involve one or more parents or substitute parents within the household, as she herself notes (1997, p. 19).

Hrdy's account was unique at the time it was proposed in the extensive use it makes of comparative evidence from nonhuman primates, and in the emphasis that it puts on the existence of variability among females in reproductive success and the components of selection involved in that success. Hrdy was also a pathbreaker in that she criticizes the pair-bond accounts of female orgasm quite harshly, emphasizing that they present the view that "women's sexuality evolved in order to 'serve' her mate" (1981, p. 176). Yet Hrdy's story is arguably also male centered, focused as it is on the dependence of the females on males for protection and help, and on the lurking danger of infanticidal males. The central image is one of prostitution; women engage in frequently unsatisfying sexual acts in exchange for aid or tolerance.

Hrdy's neglect of female-female bonds also smacks of androcentrism. Why does Hrdy not include female-female recruitment of support in her list of possible social circumstances that influence fitness? The females, as she portrays them, seem to be completely dependent for their well-being on male good will and male response. What, for instance, about the all-female coalitions of rhesus monkeys or bonobos that sometimes attack males? (Akers and Conaway 1979; or see Wrangham 1980; Irons 1983; Wasser 1983; de Waal 1995; Wrangham and Peterson 1996). Hrdy describes the general primate female as "highly competitive, socially involved, and sexually assertive," and claims that competition among females is a "major determinant of primate social organization" (1981, p. 189; cf. Caulfield 1985, p. 351). This claim is based partially on Hrdy's own valuable work on the Hanuman langurs. There are, however, species of primate closely related to us in which significant female cooperation, bonding, food sharing, and coalitions have been observed, most prominently, the bonobo. In order to support her important claim that females' actions and social systems contribute to their offspring's fitness, and are thus selectively powerful, Hrdy emphasizes the competition among females. But competition among females can exist side by side with cooperation, just as it does with males. The existence of significant cooperation among female prehumans, at least in some areas, such as defense and support, might undermine Hrdy's story, centered as it is on the dangers of male attack or neglect. Hrdy, in her recent work, acknowledges the existence of social alliances between unrelated bonobo females,

TABLE 2  Problems with evolutionary accounts in Chapters 3 and 4. Entries in parentheses indicate arguable cases, not as obvious as clear cases of mistakes, which are not in parentheses.

| Author | Female orgasm only in intercourse | Intercourse = orgasm | Conflict/no support sex research | Female response like male's | Ancestral response dictated by hormones | Conflicts with evidence in nonhuman primates | Orgasm = adaption |
|---|---|---|---|---|---|---|---|
| Morris | ✓ | | | | ✓ | ✓ | ✓ |
| Gallup and Suarez | ✓ | (✓) | ✓ | ✓ | ✓ | ✓ | ✓ |
| Pugh | ✓ | (✓) | ✓ | ✓ | | ✓ | ✓ |
| Crook | ✓ | ✓ | | | | | ✓ |
| Newton | ✓ | ✓ | | | | ✓ | ✓ |
| Eibl-Eibesfeldt | ✓ | (✓) | | | (✓) | | ✓ |
| Barash | ✓ | ✓ | | | ✓ | ✓ | ✓ |
| Campbell | ✓ | ✓ | ✓ | | ✓ | ✓ | ✓ |
| Beach | ✓ | ✓ | ✓ | | ✓ | ✓ | ✓ |
| Hamburg | ✓ | | | | | ✓ | ✓ |
| Rancour-Laferriere | ✓ | | | | ✓ | | ✓ |
| Alcock | ✓ | | ✓ | | | | ✓ |
| Allen and Lemmon | ✓ | ✓ | ✓ | | | | ✓ |
| Bernds and Barash | ✓ | | ✓ | | | | ✓ |
| Alexander and Noonan | ✓ | ✓ | ✓ | ✓ | | | ✓ |
| Diamond | ✓ | | ✓ | | | | ✓ |
| Sherfey | ✓ | ✓ | ✓ | | | | ✓ |
| Hrdy | ✓ | | ✓ | | | | ✓ |

but she does not seem to consider that such alliances might undermine her explanation of female orgasm, by alleviating some of the proposed selection pressure (1997, p. 21).

Of course, the existence of a male-centered view in an hypothesis like Hrdy's is not evidence against it, but it does raise concerns about what evidence is being left out (as she herself admits vis-à-vis the views she is criticizing). By far the most significant problem with her entire theory, however, is that neither nonhuman nor human females respond sexually in the way needed for her hypothesis regarding accumulated time of sexual intercourse. Thus, strangely, the only explicitly feminist account given of women's orgasm is shattered on the rocks of evidence about female sexuality.

## Conclusions

I have documented numerous problems with the evidence and logic of the early evolutionary explanations of female orgasm in the last two chapters. First of all, they all assume, without any evidence, that female orgasm is a special adaptation designed to increase the reproductive success of the female. As George C. Williams emphasized, adaptation is an onerous concept, one that should be applied only when there is sufficient evidence of good design (1966). The practice of explaining orgasm only as it relates to intercourse violates this standard stricture about adaptive explanations. Orgasm with intercourse is a problematic subset of female orgasm in itself. When explanations are given for female orgasm only during intercourse, they are not addressing the phenomenon they purport to explain: the existence of female orgasm itself. This fact is usually obscured by the erroneous assumption that intercourse ensures orgasm for the female. Most of the evolutionary accounts neglect the orgasm/intercourse discrepancy, and this omission constitutes a fundamental evidential problem, because a basic assumption of their explanations is false.

Then there are the specific evidential problems. I have documented ways in which most of the explanations are in conflict with contemporary sex research on female orgasm. Sometimes this conflict comes in the form of assuming that female sexual response is like the male's, when it is not. There are also the conflicts with the nonhuman primate evidence, both about the existence of orgasm in these animals and about their dependence on hormonal cues. In particular, a number of the researchers assume that the hominid and prehominid ancestors of modern women were completely dependent on their hormonal systems for their sexuality, an assumption that is violated by what is known about nonhuman primate sexuality. Note that I am applying completely usual standards of evidence here. When an explanation's basic assumptions are undermined by available evidence, then the explanation itself is undermined.

There are also the logical problems I documented. For clarity, I have assembled Table 2, which charts out which evidential or theoretical problems were manifested in the various explanations covered in Chapters 3 and 4. I conclude that there are no viable explanations left from the 18 that I considered in these chapters.

CHAPTER 5

# The Byproduct Account

In the last two chapters I found wanting 18 of the adaptive accounts of female orgasm presented in the literature. Readers may be anxious to see what a nonadaptive account—the variety I favor—actually looks like. The goal of this chapter is to introduce one such account.

In 1979 Donald Symons advanced an important theory of the evolution of female orgasm. Instead of viewing female orgasm as an adaptation, Symons proposed that "human female orgasm is best regarded as a potential" (1979, p. 89). On his view, female orgasm is a potential or capacity that is present in all mammals, but is activated in the females of only a few species: "Humans differ from other mammals primarily in that, among some peoples, techniques of foreplay and intercourse provide sufficiently intense and uninterrupted stimulation for females to orgasm" (1979, p. 89). In this chapter I explicate and defend Symons's hypothesis.

Why do human beings and other mammalian females have this potential? According to Symons, it is a byproduct of embryological development. Take human beings. As a fertilized egg grows into a fetus, it passes through a series of stages of growth. In the early stages of human embryo development, both male and female embryos have the same physical characteristics, with the exception of having different chromosomes—they are not differentiated by any external sexual characteristics. This stage continues until the male embryo experiences a release of hormones into its body, at which time the embryo starts to develop different sexual apparatus from the basic female form. If no new hormones are circulated, then the embryo develops into a girl. This spurt of embryonic hormones occurs after week 8 of gestation; until then, the female and male embryos are indistinguishable except at the level of chromosomes (Hamburg 1978a, p. 170).

It is crucial to note that the penis and the clitoris are the "same" organ in men and women; there is an organ in the primordial, undifferentiated embryo that turns into a penis if it receives a dose of particular hormones; otherwise it matures into a clitoris. In other words, the penis and the clitoris have the same embryological origins and are thus called "homologous" organs. Similarly, the nervous and erectile tissues involved in orgasm in both sexes arose from a common embryological source (Kinsey et al. 1953, pp. 571–572). Thus the kinds of tissues involved in orgasm in males and females are the same. These include: nerve tissues involved in sensing stimulation and excitement; erectile tissues, which are spongelike tissues that can become engorged with blood and stretched during sexual excitement; and muscle fibers, which are distributed in various locations in the pelvic floor of both sexes and are involved in orgasmic contractions. These tissues are what the sexual organs are built from, especially the penis in males and the clitoris in females.

There are other reasons to believe that the erectile and nervous tissue involved in both male and female sexual excitement and orgasm has a common embryological origin. For one thing, there appears to be a common neurological foundation for the reflex stimulating the muscular contractions of orgasm in each sex (Kinsey et al. 1953, pp. 571–572). The time between the beginning orgasmic contractions in each sex is 0.8 seconds. Moreover, there are profound similarities between the orgasms of women and those of prepubescent boys. Prepubescent boys, by definition, do not experience ejaculation, and some are capable of experiencing a sequence of or-gasms in a row with little or no refractory (resting) period, in marked contrast to ejaculatory males (Kinsey et al. 1948, pp. 158–159). Adult men engaging in Asian sexual practices such as Tantric yoga are capable of disengaging orgasm from ejaculation, so that they, too, are capable of experiencing repeated orgasms with lit-tle or no refractory period (Kinsey et al. 1948, pp. 158–159; Fox 1993, pp. 23–25). The key difference in refractory time seems to be whether or not ejaculation has occurred. There is some support for this conclusion in women as well. Those women capable of ejacu-lating with orgasm tend to view the ejaculatory orgasm as the final one that they want; the usual female capacity of having further or-gasms with little or no refractory period seems not to apply in the case of female ejaculatory orgasms (Ladas et al. 1982; see note 2, Chapter 2).[1] The male and female orgasms are remarkably similar.

We can gain two insights from this information about the me-chanics of orgasm. The first is that there seem to be at least two stages in the maturation of the adult male orgasmic ability: one state in which orgasmic contractions are fully wired up with the erectile tissues and the contractile muscles; and another in which ejaculation and sperm delivery are hooked into the orgasmic con-tractions. The second insight is that the first of these stages is com-mon to males and females of all ages. A concert of interactions is involved in producing orgasm in males—these interactions are pres-ent in both mature and immature males—and does seem to be par-alleled in females. Similar erectile and nervous tissue is involved in orgasm in both sexes, and the reflex itself appears to be identical in ejaculatory and nonejaculatory males and in females.[2]

There is a significant difference, though, according to Symons, between male and female orgasm, and it lies in the trait's past con-tribution to reproductive success. Symons chooses the example of the male nipple to make his point. As mentioned in Chapter 1, nip-ples are necessary to the reproductive success of any female mammal. (Milk substitutes were not available to our ancestors before the advent of agriculture, and thus could not have played a role in shaping the basic human form.) Therefore, there is strong and continuing selection pressure for female nipples. The male mammal gets nipples through sharing the same embryological form with female mammals. Thus his nipples are a byproduct of selection on the female mammal.

Symons claims that female orgasm evolved in a similar way. Orgasm and ejaculation are strongly selected in men since they use the contractile pulses of orgasm as a sperm-delivery system (Bancroft 1989, p. 86). Strong selection on the male sexual tissues for performance of orgasm and sperm delivery is hence ongoing. Just as in the male nipple case, the opposite sex acquires the equipment in virtue of an early embryological commitment. Females get the erectile and nervous tissue necessary for orgasm in virtue of the strong, ongoing selective pressure on males for the sperm delivery system of male orgasm and ejaculation. To continue the parallel, I would add that either embryological bonus—the male nipple or the female orgasmic equipment—can be used by the gifted parties; males often inherit not only the nipple structure but also the pleasurable and sexual sensitivity of the female nipple, and they can make use of this in their sexual practices.[3] Similarly, females inherit the clitoral organ and the structural erectile tissues and neural pathways needed to experience orgasm and can make use of them in their sexual practices.

### How the Byproduct Account Fits with the Evidence

At this point, it is important to review some of the apparent mysteries of female sexual response. Symons cites these data in support of his theory. In fact, under his theory we can even expect some aspects of female sexual response. Let me review some of the established data, taking as my hypothesis that female sexual morphology and the capability for orgasmic performance arising from that morphology is a byproduct of the primordial embryological development of male sexual response.

So, what would we expect to be the case if female orgasm arose as a developmental byproduct of selection on male orgasm? One thing this hypothesis would explain is the apparently strange data on female masturbation techniques. Recall that Gebhard and colleagues note that the "most common masturbation technique is the manual stimulation of the clitoris and the small lips of the vulva," which accounts for 84% of all acts of masturbation among the women the Kinsey team surveyed (Gebhard et al. 1970, p. 15). Less than one fifth of women masturbate by inserting an object or fingers into the vagina, and nearly all of those who do accompany the action with clitoral stimulation (Gebhard 1970, p. 16; Kinsey et al. 1953). Recall also that, as Kinsey himself noted, women almost never masturbate solely by inserting something into the vagina, in imitation of the act of intercourse (Kinsey et al. 1953, p. 163). And there are other independent studies that agree with these findings. For example, Hite found that only 1.5% of women masturbate by vaginal insertion alone (1976, p. 411). Moreover, women's preferences for clitoral and labial stimulation are widely known; Kinsey cites 16 sources in European and American literature, dating from 1885 (1953, p. 158).

As we have seen, these data are especially difficult to account for on an evolutionary view that female orgasm evolved in some relation to heterosexual intercourse. However, that almost no women masturbate by simulating the act of intercourse is quite comprehensible on the byproduct view because there is no theoretical commitment to holding that the clitoris's function is somehow to be stimulated during intercourse. By the same token, if framed within the hypothesis of female orgasm as an embryological byproduct, this reasoning also allows us to make sense of the otherwise puzzling data on the relative infrequency with which women experience orgasm with intercourse. Under the common assumption that the capacity for orgasm is designed as an adaptation to encourage and reward intercourse, this infrequency must be seen as a design flaw.

Perhaps adaptationists will reply, however, that I am misrepresenting the degree to which female orgasm and intercourse typically go together. Admittedly, the statistics from the sex research literature do vary, but among those studies that find a higher correlation between orgasm and intercourse, there is at least one recurring, confounding factor: these studies often do not state whether they are counting all orgasms with intercourse, or only those unassisted by direct manual clitoral stimulation. This is important because such stimulation provides a profoundly higher chance that the female will have an orgasm during an episode of intercourse, and though there is no study, per se, of how widespread the practice of assisting intercourse is, the cross-cultural data indicate that it is not widely practiced around the world, however widespread it might be in the United States or Europe (Davenport 1977).

For example, in the Hunt (1974) data on 700 married, white, American women, 53% had orgasm all or almost all of the time with intercourse, while 7% had orgasm none or almost none of the time with intercourse. But this study does not make clear whether women who had orgasm during the same sexual bout as intercourse, though not during and as a direct result of intercourse itself, were counted as having orgasm "with intercourse." If this was the case, then the study could have included female orgasm by any means before, during, or after intercourse. Indeed, Hunt's numbers suggest that he was counting some sort of assisted intercourse: the numbers for orgasm all of the time with unassisted intercourse generally fall around the 15% to 35% range (Terman 1938; Chesser 1956; Tavris and Sadd 1977; Hite 1976; Fisher 1973), while the reported percentages of women who reliably have orgasm with intercourse, *both* assisted and unassisted, range from 31% to 54%

(Kinsey et al. 1953, pp. 379–380, 382). Note that even assisted intercourse barely achieves the level of prompting orgasm more than half of the time in the average woman. Also note that the 15% to 35% of reliable orgasm with unassisted intercourse is quite low, and that it is repeatedly produced across studies with different methods and different populations (although all the women were American or European).

At any rate, few American studies of the incidence of orgasm with intercourse distinguish between assisted and unassisted intercourse. Kinsey and colleagues, for example, include assisted intercourse in their numbers for orgasm with intercourse, noting that the stimulation provided by petting and masturbation are much more suited than intercourse to inducing female orgasm (Kinsey et al. 1953, p. 391). In Fisher's study, 35% of women experiencing orgasm with intercourse did so through direct clitoral stimulation (1973, p. 193), and a full 63% of these women have orgasm through having their clitorises directly stimulated before intromission. Given the purported predominance of the sexual act that proceeds with the sole aim of male orgasm across cultures, it is unlikely that assisted orgasm is the norm (Davenport 1977).

Thus there is a low percentage of female orgasm with unassisted intercourse, and as we have seen, under Symons's developmental hypothesis, this is not surprising. Unassisted intercourse provides mostly indirect stimulation to the penis's homologue, the clitoris. The only copulatory position in which the clitoris can easily be directly stimulated is with the woman in the superior position (Masters and Johnson 1965). Thus the fact that several evolutionary writers find surprising—the low rate of reliable female orgasms with intercourse—becomes quite understandable under Symons's developmental hypothesis.

It is no surprise that, as we saw in Chapters 3 and 4, many of the evolutionary writers did not acknowledge that the rate of female orgasm with intercourse was so low, although several of them mentioned the "difficulty" or slowness of women in achieving orgasm with intercourse. Some of these beliefs may have been based on popular lore, but it is clear that Masters and Johnson's account of the mechanics of female orgasm with intercourse had a strong influence, especially upon Morris and the many that followed his account, as well as upon Sherfey. The problem with such accounts is that they treat women who do not have orgasm with intercourse as, in Masters and Johnson's word, "dysfunctional." On this view, the nearly 50% of women who, according to the sexology studies, do not have orgasm with intercourse even half the time, must not be behaving "naturally." Thus Symons's byproduct hypothesis regarding female orgasm has the virtue that the facts of female sexual response do not need to be distorted in order to make sense of it. But what could have convinced Masters and Johnson to promulgate such a distortion? Symons has an insightful answer. He writes, "The implication frequently drawn from Masters and Johnson's writing, an implication that I believe was intended, is that the female genitals are designed (presumably by natural selection) to generate orgasm during heterosexual copulation" (1979, p. 87). The first and most significant of the biases he emphasizes in Masters and Johnson's work is that their experimental subjects are not representative of women, but were required to have orgasm easily with intercourse in order to participate in the study, as I noted in Chapter 3. As Symons writes, the consequence of this sampling bias was that "since all the participating women orgasmed during intercourse, orgasm was made to seem a 'natural' concomitant of intercourse" (1979, p. 87). An additional bias, according to Symons, lies in Masters and Johnson's pro-marriage politics, which resulted in "strained attempts to demonstrate the complementary nature of male and female sexuality" (Symons 1979, p. 87; this bias is also discussed by Robinson 1976, pp. 158–159). Symons concludes, "I believe that one result of Masters and Johnson's marital bias is their implication that male and female genitals are not only complementary in their proportions but equally adapted to orgasm production during (marital) intercourse" (1979, p. 87).

## Cross-Cultural Evidence

Still, some writers harbor suspicions that the orgasm rate among nonmodern American women is higher than that shown in the American studies: in other words, that low orgasm/intercourse rates are pathological, but that the pathology is due to our modern (American) culture (Allen and Lemmon 1981). It is therefore important to consider cross-cultural evidence. Unfortunately, there is little reliably collected cross-cultural evidence; nevertheless, consider a summary of the ethnographic literature given by Symons's main source for cross-cultural information, William H. Davenport: "In most of the societies for which there are data, it is reported that men take the initiative and, without extended foreplay, proceed vigorously toward climax without much regard for achieving synchrony with the women's orgasm. Again and again, there are reports that coitus is primarily completed in terms of the man's passions and pleasures, with scant attention paid to the woman's response. If women do experience orgasm, they do so passively" _accidentally_ (1977, p. 149; quoted in Symons 1979, p. 86). Kinsey and colleagues come to a similar conclusion about American sexual practices in the mid twentieth century.[4]

Some other cross-cultural evidence is more difficult to assess. Donald Marshall and Robert Suggs present a number of studies of various ethnographic groups in their landmark cross-cultural book, _Human Sexual Behavior_. These include an account by Marshall about Mangaia in central Polynesia, where female orgasm is "universally achieved" (1971, p. 122). All of Marshall's information is from self-reporting, and it is unclear how much of it involved self-reports from the females. Hence, it is difficult to discern exactly what we should conclude from such reports. Suggs says that on the French Polynesian islands of the Marquesas "there is little sexual foreplay, and the orgasm for both parties is achieved easily and rapidly, relative to western standards" (1971, p. 168). Once again, this is self-reporting against a background of a strong cultural ideal of mutual orgasm, and it is difficult to assess the conclusion.

In any case, the overall conclusions of the volume emphasize the learned aspect of female sexual response: "The preceding chapters indicate quite clearly that female ability to respond to sexual stimulation is not necessarily inherent in female physiology but, in large measure, is learned behavior" (Suggs and Marshall 1971, p. 241). There is nothing in the Marshall and Suggs volume to contradict Davenport's claims, and Davenport's study remains the only available large cross-cultural study.

One 1981 study of West German students found rates of orgasm with intercourse more or less identical with those found in the U.S. studies. In nonmarital coitus, the survey of 528 women showed 19% never having orgasm with intercourse, while 23% almost always did. The numbers for marital coitus were similar, with 16% never having orgasm and 25% almost always having it with intercourse (Clement et al. 1984, p. 114).

More recent cross-cultural evidence also seems to support Davenport's conclusions. In a 2000 survey of college-educated Pakistani men, only 42% of them believed that women were capable of experiencing orgasm (Qidwai 2000, p. 74). In a 1990 study of Chinese medical students, only half of the 52 males surveyed agreed that "female frigidity is often wrongly self-attributed," while 77% of the females agreed (Chan 1990, p. 76). Orgasm rates with intercourse were, however, unavailable. Much more information about other cultures is needed before we can come to any firm conclusions.

## Nonhuman Evidence

There is another entire area of investigation that is relevant to
Symons's hypothesis. If he is right that female orgasm is a potential
in all female mammals, then how does the nonhuman evidence bear
on his theory? I shall confine myself to discussion of the literature
on nonhuman primates, since these are our closest relatives. In
brief, Symons concludes that evidence of female orgasm among
nonhuman primates occurs only outside of copulation, and that
such orgasmic behavior has been observed only in captivity. This
leads Symons to conclude that nonhuman primate orgasm is most
probably insignificant in the natural world.

Rather than reviewing only the evidence discussed by Symons, I
will offer an update of the literature on nonhuman primate female
orgasm. There is now evidence that female orgasm can occur during copulation among stumptail macaques in captivity, and in the
wild among Japanese macaques (Wolfe 1984). Even though I find
Symons's conclusions about the nonhuman primate evidence to be
too conservative, I find that his general thesis is supported by the
evidence: it does seem that female orgasm is a potential in at least
some nonhuman primates, and it seems clear that some of these females have learned to activate this potential for themselves.

The best evidence comes from stumptail macaques. In a set of
landmark studies, Suzanne Chevalier-Skolnikoff documented some
homosexual encounters among females in which the mounting female (the female rubbing her genitals against the back of another
female, the "mountee") displayed all of the physical features of the
orgasmic response in the males. These unique features include "a
pause followed by muscular body spasms accompanied by the characteristic frowning round-mouthed stare expression and the rhythmic expiration vocalization" (Chevalier-Skolnikoff 1974, p. 109).
She notes that all the obvious female orgasms occurred while a female was mounting another female, and not during heterosexual

copulation (Chevalier-Skolnikoff 1976, p. 521). Interestingly, the mounting females were never in estrus, nor did they come into estrus within 2 days of a homosexual encounter. Thus it appears that female orgasm in this species is independent of the estrus cycle: "since the females who initiated homosexual activities apparently were not in the ovulatory phases of their cycles, their homosexual behavior evidently was not elicited directly by estrous hormonal states" (Chevalier-Skolnikoff 1976, p. 524).

Chevalier-Skolnikoff's studies were followed up by a series of experiments performed by D. A. Goldfoot and his group. They found that in homosexual mounting episodes intense uterine contractions and sudden increases in heart rate coincided with the exhibition of the ejaculation face and panting grunts noted by Chevalier-Skolnikoff (Goldfoot et al. 1980, p. 1477). During the first study, from 1975, Goldfoot and colleagues found no ejaculation face at any time among the females while they were being tested during heterosexual copulation (1975, p. 555). In a later study, the distinctive uterine contractions and increased heart rate were also observed in 4 out of the 10 females tested during *heterosexual* copulations, and in these 4 they occurred 5% to 40% of the time. All of these findings were obtained through radiotelemetry, which amounts to wiring up the monkey with transmitter electrodes in order to obtain on-site measurements of uterine contractions and heart rates. They found a unique uterine contraction pattern upon the occasions when the climax face was shown (1980, p. 1478).

Slob and van der Werff ten Bosch repeated this type of study in 1986 and 1991. They found a pattern similar to Goldfoot's in their measurements of female homosexual mounts in which the mounter showed the climax face (1991, p. 144). They found a series of seven clonic contractions with a mean of 0.7 seconds between peaks. The uterus returned to a baseline muscle contraction level within 10 seconds after the female ended the climax face. The first 10 seconds of uterine contractions were accompanied by a sharp increase in heart rate. This pattern of responses differs markedly from the responses they recorded during heterosexual copulation, in which there was a uterine contraction associated with male ejaculation, accompanied by increased heart rate. It is especially noteworthy that the two uterine contraction patterns differ. Nevertheless, Slob and van der Werff ten Bosch argue, somewhat desperately, that female orgasm occurs during heterosexual copulation as well, even without the climax face (1991, pp. 142–144).

This move is reminiscent of the special pleading by Chevalier-Skolnikoff in her discussion of the fact that she had not observed the climax face on the females during heterosexual copulation. She remarks, "Interestingly, all the obvious female orgasms observed here occurred during homosexual interactions. However, there is evidence that orgasms may also occur during heterosexual coitus" (1976, p. 521). I shall present her reasoning in some detail, because it is a good example of the influence of bias on the process of arriving at scientific conclusions.

First, Chevalier-Skolnikoff cites Masters and Johnson's result that the intensity of human female orgasm is variable, with mild orgasms being "hardly distinguishable behaviorally or physiologically." She also notes that more intense clitoral stimulation tends to produce a more intense orgasmic response than vaginal coitus in women (1974, p. 113). Then she argues, "In view of this variability in the human female orgasm, it is conceivable that in stumptail females less intense and less easily identifiable orgasms than those observed during homosexual interaction might occur during heterosexual copulation" (1974, p. 113). Jeannette Hanby takes a similar line, claiming that "orgasm could certainly occur in some females, even though they do not show a pattern resembling male ejaculations. If orgasm is as variable and subtle—and often absent—in the non-human female as it is in the human female, it would be very difficult to observe at all, let alone reliably" (1976, p. 49).

Having established the "conceivability" of orgasm, Chevalier-

Skolnikoff offers two "indications" that stumptail females do have orgasm during coitus. First, they demonstrate reaching back and clutching behavior, which resembles the spasmodic hand grasps noted by Masters and Johnson in human females during orgasm. However, the clutching reaction is a weak candidate as a sign of orgasm. Doris Zumpe and Richard Michael describe this behavior in the female rhesus: "the female turned her head round, backwards and upwards, to look at the face of the male. This was accompanied by vigorous lip-smacking and a reaching back with one hand to grab and pull at the hair of the male's head, shoulder, lower abdomen, thigh or leg" (1968, p. 119). They claimed that this suite of behaviors usually occurred at the moment of the male's ejaculation. Linda D. Wolfe observed that Japanese macaques also displayed a clutch reaction. Cyril Fox and Beatrice Fox, however, observed that the clutching reaction in human beings "may be independent of orgasm" (1971, p. 333; cf. Slob and van der Werff ten Bosch 1991). In addition, Hanby notes that among Japanese macaques "females of any age, males, and copulating or playing pairs would often reach back more as a mount or dismount gesture than as a gesture that could be reliably associated with the end of a series of mounts or male ejaculation" (1976, p. 49). Given the occurrence of clutching behavior in a variety of circumstances, it is unlikely that it indicates orgasm. Moreover, no correspondence between the clutching reaction and uterine contractions was found in Goldfoot's studies. Hence, Symons's conclusion, that the clutching reaction is most likely a sign of female sexual excitement—but not of orgasm—is supported.

Chevalier-Skolnikoff also offers a second piece of evidentiary support for female orgasm in stumptails. She notes that the post-ejaculatory phase indicates that a genital lock probably occurs: "in other mammals [such as the dog] in which genital locks occur, they are caused by both enlargement of the penis within the vagina and simultaneous constriction of the muscles of the vagina. Masters and Johnson have noted that in the human female, vaginal muscular contractions occur during the orgasmic phase of coitus . . . It is likely that the vaginal muscular spasms that evidently occur in female stumptail monkeys—and other mammals which tie [have "locking" genitals]—are manifestations of orgasm. Thus it is likely that female stumptail monkeys also experience orgasm during heterosexual coitus" (1974, p. 113).

There is a serious problem with this second point. If we are following the human model, then before orgasm, the tissues surrounding the vaginal canal are expected to become engorged with blood, which makes the canal itself a tight fit. Orgasm serves to relieve the congestion of the area surrounding the vagina, thus opening the vaginal canal. Immediately following orgasm, then, the vagina is actually less tight than it was before. If humans are to be taken as a model of sexual response for stumptails, as Chevalier-Skolnikoff argues, then her reasoning fails; it is highly unlikely that the muscular spasms of orgasm can work to sustain a genital lock immediately following orgasm.

In spite of the weakness of these arguments and the glaring absence, during copulation, of the very distinctive suite of behaviors indicating orgasm in the females she observed, Chevalier-Skolnikoff's conclusion gets even stronger: "the unmistakable observation of orgasm in female stumptail monkeys during homosexual interactions and *strong evidence* for the occurrence of female orgasm in this species during heterosexual coitus . . . suggest that females of at least some [species of macaques] experience orgasm" (1974, p. 113; my emphasis). The curious thing about Chevalier-Skolnikoff's conclusions is that there seem to be two types of orgasm here, one in which the female exhibits the complex suite of behaviors like the males, and one in which there is a clutch reaction. Her argument here depends completely on the variability of orgasm in women, yet although Masters and Johnson found variability in female orgasm, they also documented the physical signs common to all female orgasms, for instance, body rigidity and muscular contractions of the orgasmic platform (1966, pp. 128–129).[5] If Chevalier-Skolnikoff is right, and the orgasm during coitus is a weaker version of the orgasm during homosexual mounting, why do the mountees in heterosexual coitus exhibit *none* of the distinctive behaviors of orgasm? We are to suppose that in contrast to the common core of behaviors found all along the continuum of intensities of human orgasms, there are two entirely distinct sets of behaviors correlated with stumptail orgasms.

Worst of all, Chevalier-Skolnikoff's argument for the common occurrence of a different type of orgasm with intercourse in stumptail macaques was later undermined by Goldfoot and colleagues (1980). They found that during stumptail heterosexual coitus all occurrences of the distinctive round-mouthed expression were accompanied by a distinctive pattern of uterine contractions, and vice versa.

The explanation for Chevalier-Skolnikoff's presentation of weak arguments and lack of observational data as "strong evidence" might be revealed in the conclusion she draws: "it is also likely that females as well as males experience orgasm during heterosexual coitus. This suggests that male and female sexual behavior is more than *potentially* similar, that in fact the two coital patterns *are* more similar than previously thought" (1974, p. 113; her emphasis). In other words, both males and females supposedly experience orgasm during intercourse. Her observations actually show that the sexual physiological responses of males and females to direct stimulation of the penis or clitoris are quite similar; but the social situation under which this condition was clearly satisfied for the females in her population primarily involved only other females. Why does Chevalier-Skolnikoff depart from her evidence and indulge in ill-formed hypotheses? I suggest that Chevalier-Skolnikoff is committed to showing that females get the same pleasure out of sexual intercourse that males do, regardless of her evidence. Otherwise, female orgasm would not be automatically linked to male orgasm and male sexual activity. She apparently cannot imagine that males and females have very different responses to a behavior so important to reproductive success as heterosexual copulation.

Slob and van der Werff ten Bosch (1991) give a similarly tangled data interpretation of their experiments on stumptail macaques in order to sustain the claim that female orgasm occurs very frequently with intercourse. In the Slob et al. 1986 study, they sought to explore whether the clear behavioral and uterine changes found by Goldfoot during female homosexual mounting were also found during heterosexual copulation. Both groups measured the respiration, heart rate, and uterine contractions of females during heterosexual encounters. Slob and colleagues found uterine contractions that occurred immediately after the onset of male ejaculation, as well as various uterine contractions and motions during intercourse itself. They also found that the characteristic climax face, found during homosexual mounts and some heterosexual copulations by Goldfoot and colleagues, is *not* correlated with the uterine contractions that occur immediately following ejaculation, which suggests that something different is happening besides orgasm. But they appear to assume that any uterine contraction is an indication of an orgasm, and thus, since each male ejaculation is accompanied by a pattern of uterine movements, they conclude, "Female sexual climax may occur during every copulation" (1986, p. 894). They allow that "alternatively, sexual climax may only have been reached when the climax face was displayed. This implies that the uterine movement pattern may not be a reflection of the female's climax, but be secondary to penile movements or penile effusion" (1986, p. 894). Strangely, they reject this alternative on the basis of findings from homosexual mounts by females, in which both the climax face and a uterine contraction pattern were shown. They conclude that the presence of such contractions makes it unlikely that the uterine contractions in heterosexual copulation are caused by penile movements.

The real question here is whether the heterosexual uterine contraction pattern is similar enough to the contraction pattern shown in homosexual mounts that they can both be treated as orgasms. Slob and van der Werff ten Bosch's 1991 study allows a closer comparison. There they found a characteristic post-ejaculatory pattern of uterine movements that includes an initial rise in contractional force followed by a relaxation within the next 15 seconds. This occurred without the presence of the climax face. They also recorded three copulations in which one female showed the climax face four times. During their discussion of these findings, Slob and van der Werff ten Bosch defend the idea that females are having orgasms without showing the climax face; they say that this shows in the heart rate evidence. They found that the female heart rate rose in conjunction with the male ejaculation. Ultimately, though, it is the occurrence of uterine contractions after every ejaculation that convinced them that "the uterine contractions and climax face seem to be unrelated." "Thus," they conclude, "female sexual climax may occur during every copulation, and be reflected by the early post-ejaculatory pattern of uterine movements" (1991, p. 143).

Once again, they appeal to the evidence from homosexual mounts to buttress their claim that orgasm occurs with every heterosexual copulation. Their own studies of female orgasm within homosexual mounting contradict their desired conclusion, however. They recorded two mounts that showed the climax face, and show the graph of the contraction pattern that occurs with the climax face in homosexual mounts. This graph shows an "intense . . . sustained tonic uterine contraction" beginning 8–10 seconds before overt behavioral changes and continuing for *40 to 50 seconds*. Six seconds after the major tonic contraction, there were seven clonic (pulsing) contractions with a mean of 0.7 seconds between peaks.

Thus the contractions associated with the climax face are markedly longer than the 15-second contractions that routinely show up following male ejaculation. They also involve repeated, pulsing contractions, unlike the single contraction that occurs during ejaculation. Given the wide variety of uterine movement and contraction patterns they recorded, there is no reason to think that the long, intense contraction pattern associated with the female showing the climax face and with muscle tension and erection of hairs is in any way similar to the brief uterine contraction following ejaculation that occurs without the other behavioral displays.

In sum, their own evidence belies Slob and van der Werff ten Bosch's conclusion that female stumptails experience sexual climax with every copulation. Far from supporting their conclusion, as they claim, the homosexual evidence undermines it; the pattern of response in orgasmic homosexual mounts is markedly different from what they claim is the universal incidence of female orgasm following male ejaculation. Like Chevalier-Skolnikoff, they seem to be willing to create an entirely new category of orgasm in order to sustain the claim that female orgasm occurs consistently with heterosexual copulation.

Both Chevalier-Skolnikoff and Slob and van der Werff ten Bosch clearly interpret their results on the assumption that it would make no sense to have female stumptails have orgasms with each other and not during heterosexual copulation. The fact that Goldfoot and colleagues (1980) did record some clear orgasms during heterosexual copulation is perhaps not comfort enough—the rates were nonexistent among 6 of the 10 females, with 4 females showing orgasm at least once with heterosexual intercourse, and 1 female clocking more than half of the recorded orgasms. With this outlier removed, the other three orgasmic females out of the group of 10 evidenced orgasm a sum total of 18 times out of 156 copulations. Still, Goldfoot and colleagues have demonstrated that some stumptail macaque females can experience orgasm with heterosexual intercourse; it just seems to be a highly variable trait both within and across individuals. In fact, the gist of Goldfoot and colleagues' data is not different in upshot from the human data; some female humans and macaques are capable of having orgasm with intercourse a fair amount of the time, while some vary, and some do not have orgasm at all with intercourse.

Other evidence exists for orgasm in several species. For example, Frances Burton performed a series of experiments on rhesus macaques that provided evidence that females of the species are physiologically capable of experiencing orgasm. Her experiment on three adult female rhesus macaques consisted of placing them in a metal framework with harnesses, grooming them, and then providing 5 minutes of clitoral stimulation, 5 minutes of vaginal stimulation, followed by a 4-minute rest, and then 5 minutes more of vaginal stimulation. Burton notes, "While arbitrary, these time units were chosen to well exceed the duration of an episode of actual stimulation and intromission in a copulatory series in the natural state" (1971, p. 182). Two of the females were tested seven times, one only four times. Each animal was tested from the 4 days prior to her putative ovulation to 2 days into the follicular phase of her cycle "to maximize her responsiveness" (1971, p. 182).

Burton's findings indicate that the monkeys "clearly exhibited" three of Masters and Johnson's four stages of the sexual cycle, which include excitement, plateau, orgasm, and resolution. She concludes that it is "not absolutely clear" whether the females had orgasms, but cites the importance of two important behavioral phenomena. One of the females exhibited spasmodic jerking of the arms and body, while another "had a series of intense vaginal spasms, to a maximum of 5 contractions which also involved the anus on 4 occasions" (1971, p. 185). The female who had the spasmodic arm jerking also had vaginal contractions, but they were not as intense as in the other female. Burton emphasizes that in women regular vaginal and anal spasming is found *only* during orgasm (she cites Masters and Johnson, 1965, p. 81; Burton 1971, p. 185). Burton is very cautious in her conclusions, however, and claims that the presence of the spasms is worth investigating, but does not prove the presence of orgasms in rhesus females. The conclusion that Burton has, in fact, documented female orgasm in this species is supported by Michael and colleagues (1974), who observed tremors in the thigh and tail muscles of a female rhesus who was mounting and rubbing herself against a male rhesus.

Regarding the occurrence of orgasm under normal copulatory circumstances, Burton emphasizes that the length of time of stimulation is critical for the (probably) orgasmic response and that "frequency of intervals within a copulatory series, and the time of individual copulations[,] probably do not permit sufficient stimulation for achievement of orgasm" (1971, p. 186). In summarizing her findings, Burton concludes, "although [female rhesus macaques are] assumedly capable of orgasm, the short duration of a copulatory series in the wild suggests that this behaviour does not occur in the natural state" (1971, p. 189).

There is one particularly interesting aspect of Burton's findings; she noted that during the beginning of the plateau phase, in which the vaginal barrel widens and deepens with vaginal stimulation, "the animal may . . . reach back with one hand to the experimenter, clutching at the experimenter's hand, or towards her own genitals. She may look back at the experimenter and she may utter a low grunting sound" (1971, p. 184). Burton links these behaviors to the clutching reaction; the important point is that these behaviors occur not during the vaginal spasms, but during the plateau phase, before orgasm. This fact implies that although the clutching reaction may be evidence for sexual excitement, as Symons suggests, we have reason to believe that it is not an indicator of orgasm.

Observations of the Japanese macaques provide evidence that a third species of macaque is also capable of orgasm. Wolfe observed female Japanese macaques masturbating, which resulted in rhythmic contractions in the muscles surrounding the genitals and anus (1991). Again, orgasm is the only known mechanism that induces rhythmic muscular spasms in these muscles.

It is also likely that female orgasm occurs in chimpanzees and bonobos. Allen (1977) manually stimulated the clitoris and vagina of a female chimpanzee, which resulted in rhythmic vaginal contractions, limb spasm, clitoral tumescence, and other indicators of orgasm. Female chimpanzees masturbate sometimes, and might experience orgasm that way (Temerlin 1975; Goodall 1986). Female bonobos may also experience orgasm in the wild (Kano 1992, pp. 144–145). Female bonobos frequently engage in a special behavior called GG rubbing, in which they rub their genitals, and especially their clitorises, against each other. The primatologist Richard Wrangham states that sometimes female bonobos reach orgasm during such encounters in the wild, as can be verified from the muscle spasms and abrupt pause (personal communication). The primatologist Frans de Waal also finds that female orgasm sometimes occurs among captive bonobos (personal communication). In sum, female orgasm has been observed in several monkey and ape species.

I should mention at this point that Symons himself is underwhelmed by the evidence for female orgasm from nonhuman primates. At the time he was writing, only Burton's, Michael's, Chevalier-Skolnikoff's, and Goldfoot and colleagues' evidence was in. Symons concludes that much of the purported evidence for female orgasm in nonhuman primates is best interpreted as showing female sexual excitement and not orgasm (1979, p. 81). Regarding Burton's 1971 tests on rhesus females, Symons emphasizes that the sort of stimulation provided by Burton would never be found in the wild, since the duration of intromission in the wild is 3 or 4 seconds. Thus, Symons concludes, rhesus "female orgasm is extremely unlikely to occur in nature" (1979, p. 81). Regarding Chevalier-Skolnikoff's stumptail data, Symons remarks that they do not show the existence of female orgasm during heterosexual copulation. Symons does accept both Chevalier-Skolnikoff's evidence in stumptails and Michael and colleagues' (1974) evidence in rhesus macaques that these females can experience orgasm when they mount other animals. However, he notes, these events were all observed during captivity, and are not supported by observations from the wild (1979, p. 82).

Symons's conclusions about the existence of female orgasm with intercourse ought to be revised in light of the evidence from Goldfoot and colleagues and Slob and van der Werff ten Bosch regarding the occurrence of the climax face on female stumptails during heterosexual copulation. Nevertheless, it still seems fair to conclude that the incidence among nonhuman primates of female orgasm with heterosexual intercourse is rare, especially in the wild. Most important, the nonhuman primate evidence seems to comport with what we would expect if female orgasm is not itself an adaptation, but rather a byproduct of development and selection on male sexual performance. It is highly variable across females, appearing in some and not in many others, and does not appear to be related to reproductive sex. Therefore, female orgasm seems to be a feature that some female monkeys and apes can learn to take advantage of for their own pleasure, which is what one would expect under Symons's hypothesis.

Although Symons does not mention the subject at all, we may wonder what it is that makes stumptail and human females so much more likely to learn to attain orgasm than females of other species. Following the logic of the developmental account, one would expect that species in which the males are under especially strong selection for male performance would be the same species in which the relevant tissues in the females would be especially well developed and susceptible to use. In other words, if the male sexuality of a particular species was especially emphasized during evolution by selection, we would expect the females to be more likely to be capable of having orgasm. This is a refinement that Symons does not consider; his hypothesis is much more sweeping: that all female mammals have the capacity for orgasm. But I think this view is too broad. If the byproduct account is right, then those species in which the sexual organs and tissues are most intensely selected in the males would also be expected to have highly sexual—and perhaps orgasmic—females.

Some support for my inference might come from the data on stumptail macaques. Male stumptail macaques have been recorded as experiencing as many as 59 ejaculations during 6 hours (Slob and van der Werff ten Bosch 1991, p. 136). This is a much higher rate than that found in other macaque species, where males ejaculate less than 10 times per day. Hence, there may be an especially well-developed sexual response system in stumptails that the females inherit developmentally that makes them especially prone to orgasmic response. Thus there is suggestive evidence from the males of the stumptail macaque species that highly sexed males evolve along with potentially orgasmic females. The male chimpanzee and bonobo are also considered highly sexed, and there is evidence that both female bonobos and female chimpanzees are capable of orgasm. Obviously, human males cannot reach high ejaculatory rates, but there might be yet unknown selection pressures on human male performance. One striking difference between human and most other ape males is the length of copulation time—a matter of minutes versus a matter of seconds—and this may be a sign of selection pressure. Orangutans are the other great ape that has extended intromission time, averaging 10 minutes (Kano 1992, p. 156). Perhaps the vast differences in timing might be a sign of an as yet unknown selection pressure. This is clearly speculative, and much more research is necessary into the correlation between male and female sexual performance in a variety of species. The evidence presently available is only suggestive.

One very important point should be noted before I continue.

Symons is not arguing that the clitoris and other tissues have no evolutionary role; his argument concerns the specific reflex of orgasm only. In fact, Symons grants a special evolutionary purpose for the clitoris: "The only known function of the clitoris in the great majority of mammalian species is to generate sensation—presumably pleasurable—during copulation" (1979, p. 88). In other words, there could well have been selection maintaining the existence of the clitoris and its sensitivity in the various mammalian species, because it serves a role in motivation and in facilitating the female to engage in intercourse. Such a contribution to reproductive sex could clearly be a target of selection. Thus Symons is not denying that the clitoris, as an organ of sensation, may have an evolutionary function for which it has been targeted by natural selection. Nevertheless, this reasoning does not extend to the use of these same tissues for female orgasm. Symons believes that the data on masturbation, the infrequency of orgasm with intercourse, women's statements about their sensations during intercourse, and the scarcity of evidence that nonhuman females experience orgasm during intercourse all undermine any claim that "women's genitals are designed to generate orgasm during intercourse" (1979, p. 89).

## The Evidence of Variability

Further evidence against orgasm being an adaptation in females comes from the variability of the trait itself, according to Symons. Symons defines an adaptation as "an aspect of structure, behavior, or psychology that has been produced by the operation of natural selection" (1979, p. 89). Thus female orgasm is an adaptation "only if in ancestral populations orgasmic females enjoyed greater average reproductive success than nonorgasmic females" (1979, p. 89).

Symons then argues: "Although there is, and probably always has been, enormous variation in the ease with which [female] individuals can orgasm—and hence ample grist for the evolutionary mill—there is no compelling evidence that natural selection favored females that were capable of orgasm, either in the evolution of mammals or specifically in the human lineage; nor is there evidence that the female genitals of any mammalian species have been designed by natural selection for efficiency in orgasm production" (1979, p. 89).[6] Here, he is relying on the comparative evidence from other species, rather than on anything from the paleontological record.

However, the data I have reviewed thus far bear him out. Recall the wide variety of female responses to intercourse. In the Chesser study of over 2,000 married English women, 24% always had orgasm with (assisted and unassisted) intercourse, 35% frequently had it, while 26% sometimes had orgasm with intercourse, 10% rarely had it, and 5% never did (1956, p. 423; cited in Symons 1979, p. 83). Similar variation was found by other researchers (see Table 1). Moreover, the cross-cultural evidence is a problem for any theory of female orgasm as an adaptation because it also reflects such variability (Symons 1979, p. 312). This evidence is a problem because it appears that female orgasm is, if anything, less common in women generally than in European or American women. In addition, Symons argues, natural selection could not have acted on a trait that was only a potential and was not actualized: "If, throughout most of human evolutionary history, the potentials of female sexuality were very rarely realized, these potentials would have been largely 'invisible' to natural selection, and this may account for the astonishing sexual plasticity of the human female" (1979, p. 313).

We need not go as far as Symons does when he emphasizes the rarity of female orgasm, but when we consider the fact that in some cultures, women's ability to have orgasms is highly developed, while in others, it is apparently not, we can see that the basic female human being could either be trained to experience orgasm with the appropriate stimulation, or not; it seems to depend quite a bit on her cultural context. This is what is meant by "plasticity." Thus there seems to be intercultural variability of female orgasmic capacity to a remarkable degree. Moreover, we have also seen the statistics on orgasm with intercourse that indicate great intracultural variety as well: even if there is a cultural context in which female orgasm is valued and women are led to participate in it, it occurs unreliably with intercourse. It must be remembered that somewhere between 88% to 96% of women in studies of these cultures are capable of having orgasms; they just do so at varying rates with heterosexual copulation (see Chapter 2; Kinsey et al. 1953, p. 513, cites a 91% rate with masturbation). Thus the data support the conclusion that female orgasmic response to intercourse is highly variable, and does not reflect whether or not these same women are capable of experiencing orgasm.

In addition to the high variability within women of the occurrence of orgasm with intercourse, there is a large amount of variability concerning the ease with which various women experience orgasm at all (Kinsey et al. 1953). Some women can have orgasm without any physical stimulation at all, while others require deep and prolonged stimulation of the clitoris. Thus female sexual response varies widely across women, which is exactly what would be expected if it were not under direct selection pressure.

All of this appears to support Symons's view that there has not been selection on orgasm with intercourse. Actually, the evidence supports that conclusion only under the supposition that we are talking about selection in one direction, toward an increased orgasm rate with intercourse. If we consider other selective regimes, all bets are off. But highly directional selection—in which an increased ability to have orgasms is more adaptive—is exactly what authors pushing an adaptationist interpretation of female orgasm have proposed. In other words, the reigning adaptive accounts assume that orgasm occurring during intercourse was specifically selected. If this had happened, we would expect, all things being equal, that orgasm would be well correlated with intercourse, but it is not; instead, we have a huge range of variability among orgasmic women. These data do not highlight the further variability that we would get from looking at different cultures. Thus we have a very high incidence of female orgasm (in some cultures), conjoined with a highly variable incidence of orgasm with intercourse. In addition, female orgasm needs to be correlated with reproductive success in order to consider it an adaptation (Sinervo and Basolo 1996).

Variability in a trait, such as we find with female orgasm, is, on first glance, a sign that selection on the trait is not occurring. If selection was occurring, we would find a concentration of the trait at one or the other extreme, in the case of directional selection, or at some intermediate value, in the case of balancing selection. Thus the evidence of widespread variability in the trait of female orgasm, which manifests itself in the fact that about 10% of women in American and European populations never have orgasm, seems to support a hypothesis that selection is not occurring. Similarly, the very wide variability of rates of orgasm with intercourse suggests that there is no selection on female orgasm with intercourse. If there was, then we would expect a high, consistent expression of the trait of female orgasm with intercourse, which is exactly what we do not find. If, however, we consider the cross-cultural data, it seems that the trait of female orgasm is quite plastic, that its frequency depends on its environment (in this case, sexual practices of the culture in question). The fact that female orgasm is unknown or very rare in certain populations with certain sexual practices, while it is widespread in other cultures with other sexual practices, implies that the trait is widely variable depending on the environment. This is technically known as phenotypic plasticity. "*Phenotypic plasticity* is the property of a given genotype to produce different phenotypes in response to distinct environmental conditions" (Pigliucci 2001, p. 1; his emphasis). Phenotypic plasticity usually has a genetic basis, and it can be selected. In such cases, the organism develops into the phenotype with the higher fitness in a particular environment. In the case of female orgasm, however, there is no evidence to indicate that either having the trait or not having it has any fitness consequences. Hence, while the trait is certainly plastic—its appearance varies with environment—there is no reason to think that it has been selected for plasticity.

Hence, although wide variability in a trait may be the result of selection on plasticity, there is no reason to think that such is the case with female orgasm. In cases in which wide variability occurs without the suggestion of selection on phenotypic plasticity, selection itself is considered unlikely to have taken place, because selection tends to narrow variability, either because the most fit phenotype is at one end of the range or because it is at the middle of the range. In either case, variability is reduced by selection over time. Thus without support for an account of selection on phenotypic plasticity, the variability that we see with female orgasm is evidence that selection has not acted on the trait at all.

No one has offered a reason why orgasm, as a behavior associated with specific morphological features, would not be selected in a directional way, such that it would reach fixation in the population, and all women would have orgasm with intercourse all the time. Hence, Symons's conclusion is exactly correct that the variability itself argues against the existence of a (directional) selection pressure on female orgasm, at least in the absence of contravening selection pressures.

In sum, there are a variety of expectations entailed by Symons's developmental hypothesis regarding female orgasm, all of which reveal supporting evidence for his view. The patterns of human female sexual response, particularly in masturbation and in intercourse, lend support to his view that female sexual anatomy and physiology has not been selected to favor orgasm with intercourse. In other words, the byproduct account suggests that similar stimulation to homologous organs would be required to achieve orgasm, and such stimulation of the clitoris is only sufficient in a minority of cases of intercourse to ensure orgasm. In addition, the high degree of variability in rates of female orgasm supports Symons's claim that no directional selection pressure has effectively shifted the population toward female orgasm. Cross-cultural evidence supports his claim that female orgasm is a capacity that may or may not be utilized. Finally, the nonhuman primate data indicate that female orgasm is a capacity that is sometimes activated among some species, and it seems to bear little relation to heterosexual copulation. In my judgment, Symons's hypothesis that orgasm is strongly selected in males, and that females get the capacity for orgasm through embryological development, has better evidential support than any other available hypothesis regarding female orgasm. It makes sense of a number of known facts about human female sexuality as well as nonhuman results, facts that are embarrassments to other, selective theories. Nevertheless, Symons's account has encountered a great deal of resistance. I shall review a few of the specific complaints against Symons's view in the rest of this chapter; then I shall discuss, in Chapter 6, an important theoretical issue underlying the disagreements over the evolution of female orgasm.

## Objections to the Byproduct Account

I shall review four points of contention involving the byproduct view: (1) claims regarding nonhuman primate orgasm; (2) feminist criticisms; (3) complaints about Symons's use of cross-cultural data; and (4) claims about the selection pressures involved in female orgasm.

### Nonhuman Primate Orgasm

One set of objections to Symons's theory of the evolution of female orgasm has to do with his use of the evidence from nonhuman primates. Hrdy disagrees with Symons that female orgasm is unlikely to occur in nature in nonhuman primates. Appealing to Burton's 1971 results from studies of the rhesus macaque, Hrdy claims that the large number of copulations during a single day can provide an accumulation of arousal that would be enough for the nonhuman primate female to have orgasm (1979, pp. 311–312). Still, she concludes that the nonhuman data on female orgasm "are too ambiguous to permit conclusions either way" (1979, p. 312).

Wolfe has a stronger argument against Symons's conclusions about nonhuman primate female orgasm. Symons based his rejection of nonhuman primate female orgasm on the fact that the observed cases of orgasm he considers involve "direct and prolonged stimulation of the clitoris or clitoral area, either by experimental design or by rubbing against another animal" (Symons 1979, p. 83). As Wolfe notes, Symons assumes that nonhuman primate females experience only indirect stimulation of the clitoris during intercourse, as human females do. Given the short period of intercourse and the availability of only indirect stimulation, argues Symons, nonhuman primate females would not receive enough stimulation during intercourse to produce orgasm. But, Wolfe argues, this is to assume a mistaken view about nonhuman primate female anatomy. Taking a closer look at this anatomy, we find that the clitoris is, in most nonhuman primate cases, in a position very near the opening of the vagina, and is thus available for direct stimulation during intercourse. Wolfe argues, "The difference in genital anatomy between humans and [some nonhuman primate] females suggests that the latter may find it *easier* to achieve orgasm through heterosexual intercourse than do women" (1991, p. 128; her emphasis).

In human anatomy, Wolfe explains, the urinary meatus (opening) is located in the vulva between the vagina and the clitoris, thus putting the clitoris away from the base of the vagina. This is in contrast to nonhuman primate females, where the urinary meatus is located either inside the vagina or near its base. The most likely explanation of the human case is that, as the size of the newborn's head increased over evolutionary time, there was selection to move the urinary meatus away from the vagina to protect it from tearing during childbirth (1991, p. 128). Wolfe proposes, "During hominid evolution, the nature of clitoral stimulation during heterosexual intercourse changed from some degree of direct stimulation, as it likely still is in the allocatarrhines [all Old World monkeys and apes besides human beings], to the indirect (and often inadequate) stimulation experienced by women" (1991, p. 129).

In addition, Wolfe argues against Symons's claim that copulation in nonhuman primates does not last long enough to produce female orgasm. Wolfe replies that the amount of stimulation necessary for male and female orgasms is species-specific. In chimpanzees, for example, copulation time (and time to male ejaculation) is a median of 7 seconds, while in bonobos it is 13 seconds, and 10 minutes in human beings (Wolfe 1991, p. 129). Wolfe emphasizes that we do not know enough about the relation between genital stimulation and the neurology of orgasm—including its timing—in either human beings or nonhuman primates, and such knowledge is necessary to settle the questions regarding female orgasm.

Wolfe is right that such research is desirable, but that does not mean that there is not any evidence already available. Indeed, she seems to miss Symons's use of this evidence, and even the existence of such evidence altogether. Wolfe does not seem to take account of the fact that evidence of female orgasm during heterosexual copulation among nonhuman primates is scarce. She asserts that there is evidence for orgasm in nonhuman primates, but her sources include cases of clutching behavior and other reactions now known *not* to be correlated with the muscular spasms of orgasm (1991, p. 129). And she takes no account of the fact that the best evidence for female orgasm in the stumptail macaques arose in homosexual mounts—a feature of the evidence that Symons takes very seriously. Of course, there is now evidence of orgasm in heterosexual copulation in stumptails, but Symons could not take account of data that were unavailable when he wrote. But this does not mean that Wolfe's view is supported, for, in contrast to her vision of nonhuman primates being highly orgasmic during intercourse, the relevant females were only infrequently orgasmic—recall that the most orgasmic individual had orgasm only 40% of the time with heterosexual copulation. Thus although Wolfe's suggestions for future research are good ones, and her comments about female anatomy are thought provoking, the evidence, as it now stands, does not support her contention that nonhuman primate females "find it easier to achieve orgasm through heterosexual intercourse than do women" (1991, p. 128).

*Feminist Objections*

Many of the objections to Symons's account have come from scientists and other researchers who take a feminist stance. Mina Caulfield, for example, objects to Symons's discussion of female orgasm, noting that he "consistently stresses [orgasm's] use for males" (1985, p. 348). She describes Symons's vision as involving "male control over female sexuality, coupled with an explicit denial of the significance of female sexual pleasure" (1985, p. 347). Similarly, Samuel Wasser and Mary Waterhouse claim that Symons's byproduct theory of the origin of female orgasm shows "androcentric value judgments and a predominantly male focus" (1983, p. 23). Sarah Blaffer Hrdy remarks that "a gentlemanly breeze from the nineteenth century drifts from the pages, bringing with it distinct déjà vu" (1979, p. 311).

Suzanne Frayser objects that Symons's byproduct view of female orgasm "dismisses the possibly important adaptive significance of the experience for females." "Is it not possible," she writes, "that heightened sexual stimulation (culminating in orgasm) is just as adaptive for females as for males, but in a different way?" (1985,

p. 39). According to Frayser, much of the problem lies in Symons's understanding of primate female sexuality itself; Frayser notes that both human and nonhuman primates engage in much more nonreproductive sex than most mammals do (1985, p. 41). Frayser criticizes Symons for not giving enough weight to the circumstances of reproduction; he focuses primarily on copulation and conception, and not on a broader sexual context that includes ways in which heightened sexual activity—especially orgasm or nonreproductive sexuality—could contribute to a female's reproductive success (1985, p. 41).[7] As Hrdy puts the point, "From a male point of view (and, in this instance, Symons') the answer [to the question 'what good is sexuality?'] is it promotes insemination; nonreproductive copulations are largely wasted effort" (1979, p. 312).

Thus the claim is that Symons's focus on reproductive sex skews his understanding of how female reproductive success relates to female sexuality. Julia Heiman has a related objection; she claims that Symons assumes "that all sex or orgasms occur in reproductive (intercourse) contexts and that 'natural' intercourse excludes any simultaneous manual touching of genitals. Thus information on orgasmic frequency during other types of stimulation is not discussed" (1980, p. 190). Hrdy concludes, "An overly narrow perspective on sexuality has led him to underestimate the selection pressures molding female sexuality during the course of hominid evolution" (1979, p. 313).

Such objections are seriously misplaced. Symons spends a great deal of his chapter on female orgasm discussing the *separation* of orgasm from reproductive sex. I suggest that the reactions cited above have their source in other parts of Symons's book, and not in the chapter promoting the byproduct view of female orgasm. In fact, Frayser, in discussing his views on orgasm, explicitly objects to the fact that Symons calls female sexual intercourse a "service" or "favor" that females "bestow" on or withhold from males (1985, p. 42). Symons does, in fact, present this idea in a different chapter of the book, which is provocatively titled "Copulation as a Female Service." It is easy to understand why feminists would object to characterizing female participation in intercourse as a service provided by females to males. But Symons's speculations on what motivates the behavior of intercourse in females can and should be separated from the consideration of the evolutionary origins of female orgasm itself. Having a highly provocative and thinly supported theory about one trait—the motivation for intercourse—does not detract from the plausibility or support of his logically independent theory about female orgasm. Although it might be suspected that Symons needed the conclusion that female orgasm is not itself an adaptation in order to argue that copulation is a service, even that would still not make the hypothesis about orgasm false. The evidence for the theory about the evolution of female orgasm must be considered on its own merits. Hrdy points out in her review of Symons's book that he takes an "opportunistic approach" to evidence, citing everything from fiction to surveys performed for *Playboy* to personal intuition (1979, p. 309). Symons does not deny that much of his book is "rhetoric" (Symons 1980b, p. 203). Nevertheless, the supporting evidence offered in the orgasm chapter is, as I have attempted to show in the previous section, relevant and appropriate.

Hrdy has objected—and other feminists have echoed the worry—that seeing female sexual response as embryologically derived from male sexual response somehow denigrates females (1981, pp. 165–166). She wants to claim that female orgasm has evolved as its own adaptation to female selection pressures, and views any theory in which female response depends on male response as antifeminist. It is important to see that there is nothing inherently antifeminist about the thesis being explored here. Where and how female sexuality arises is not significant compared to whether or not female sexuality is *controlled* by women or by men. Women have endured a long history of male attempts to control the expression of female sexuality, and many women still suffer under systems of abusive patriarchal control of female sexuality, such as those involving the surgically mutilating practices of infibulation and clitoridectomy (Terry 1999). The real problem with Hrdy's feminist objection is that it assumes that in order to be really important, female sexuality, and in particular female orgasm, must have been a direct target of natural selection among females. But there is no reason at all to think that only directly selected traits are "important." Many traits that are not understood to have arisen by direct selection are considered extremely culturally important, such as refined musical ability, the ability to design rockets, and even the ability to read. Similarly, even strongly selected traits, such as the swallowing reflex, are often devoid of cultural importance. In other words, I believe that much of the feminist reaction against the thesis that female orgasm is an embryological byproduct of selection on the male orgasm is based on a false equation of what is *important* with what is *naturally selected*.

Under the view taken here, describing female orgasm as a nonadaptation simply amounts to making a claim about how it came to be present, historically, in the human population. Culturally, nothing is implied by that fact. The trait can still be seen as extremely important culturally, or not. *Its historical genesis does not dictate our cultural attitudes toward female orgasm.*

Anne Fausto-Sterling, Patricia Gowaty, and Marlene Zuk recently attacked Symons's byproduct hypothesis on feminist grounds,[8] too, and appeal approvingly to David Buss's claim—one based on highly suspect data—that orgasm increases sperm retention (1997, p. 416, n. 3; this hypothesis and the quality of the data for it are the topic of Chapter 7). "In Buss's version of evolutionary psychology," they write, "women have much more agency than they do in Symons's" (1997, p. 416, n. 3). Here, they are equating human agency with a trait's status as a contributor to current fitness.[9] But there is at least one feminist who doesn't buy this equation (Haraway 1989, pp. 349–367). There is no basis for such an equation either in biology itself or in our wider understandings of biology's cultural significance.

Hence, I conclude that the feminist objections, considered above, to the byproduct account of female orgasm, are not well founded. They are understandable in light of Symons's other views, but the account of orgasm must be considered independently on the biological evidence that has been presented. It is a mistake to think that evolutionary adaptedness dictates cultural importance, and it is thus not a good reason for rejecting Symons's theory.

*Use of Cross-Cultural Data*

Jane B. Lancaster and Chet S. Lancaster object to Symons's use of the ethnographic (cultural) record of human sexual behavior, and complain about the "poor and male-biased quality of the data" (1980, p. 193). The Lancasters claim that the data from higher-density horticulturalists and agriculturalists (farmers) should not be used, because these lifestyles are known to correlate with increased male control over female sexuality. But Davenport (1977), Symons's main source of cross-cultural data, uses a sample of records from a variety of different societies, only some of which are agriculturalist. The issue of what the ethnographic records really represent is an important one. I have attempted to address some of the worries that Davenport's cross-cultural data is somehow skewed by supplementing his sources with other, more recent sources. The new cross-cultural data are only suggestive, but none of them challenge and some even bolster the support Davenport's findings give to Symons's hypothesis.

Daniel Rancour-Laferriere (1983) also has a complaint about Symons's use of Davenport's survey. Noting that Davenport concluded that sex around the world tends to proceed with male control and in furtherance of male desires, Rancour-Laferriere questions whether this really means that females do not have orgasms. Noting that there haven't been any physiological studies on the cultures that were included in the survey, Rancour-Laferriere laments that "the anthropologists have had to take women at their word" (1983, p. 323). Admittedly, asking women about orgasm is not the same as finding out what is really going on; these studies suffer the same weaknesses as all self-reports in sexology. But Rancour-Laferriere is advocating the use of a gold standard here: wire up the women to electrode recording devices and see whether they have the various physical manifestations of orgasm, including pelvic contractions and so on. Actually, this is not such an outrageous request. In their research, Hartmann and Fithian (1994) performed just such a test on 774 women, discovering a number of women who were having orgasms—by any physiological measure—but did not realize that they were. Once the orgasm was identified for these women, however, they easily discriminated further orgasms, which showed that they did *experience* orgasm; they simply did not have the right name attached to the experience.

There is also the issue of Davenport's strange claim that if women from many of these cultures do experience orgasm, they do so "passively" (1977, p. 149). What can this mean? I don't think it means that Davenport thinks there are two types of female orgasm, the active and the passive. Rather, he is apparently referring to the means by which orgasm occurs. According to his findings, there was little or no attention paid in most cultures to either female sexual excitement or female orgasm. Thus if a woman did have an orgasm with heterosexual intercourse, it would be "accidentally," or without the intent of the man. As we see from the American and European sexology studies, nearly all women who do have reliable orgasm with intercourse do so with a fair amount of foreplay and/or direct stimulation of the clitoris during intercourse (Fisher 1973). Thus in cultures in which the female orgasm during intercourse is not seen as a goal, it is unlikely that many women would experience orgasm.

Although Rancour-Laferriere is right that physiological data would be ideal for interpreting the sexual practices from other cultures, there is actually no record of a survey on orgasm frequency with respect to intercourse being done anywhere using physiological data. Masters and Johnson, for example, did not sample the population with a hard-wiring technique. The information gathered by researchers about the incidence of orgasm with intercourse all comes from the self-reports of women. Hence, I see no reason that other cultures should be held to a higher standard.

*Selection Pressures*

An important objection has been made to Symons's reasoning that in nonhuman primates female orgasms in the wild would be too variable and unpredictable to be selectively significant—to be targets of selection. Hrdy challenges this reasoning, claiming that the Skinnerian principle of *intermittent reinforcement* can be used to help explain the selective efficacy of even infrequent female orgasm (1979, p. 312; I reviewed problems with the intermittent reinforcement arguments in Chapter 4).

But there are some general points to be made. Most important, it is possible for natural selection to operate on a trait that has only a tiny increase in fitness connected to it. Thus even the smallest competitive edge for females that have orgasm would be sufficient for an adaptive account. The problem with this objection to the byproduct account is that, under this scenario, we would still expect the trait under selection to be universal, barring any counteracting selection pressure. In fact, female orgasm is very widespread in our species, regardless of its rate with intercourse, but it is not nearly universal among women. Why do approximately 1 in 10 or 15 women never have orgasm at all? Even with a small selection pressure pushing in the direction of an adaptive value for orgasm, we would still expect the population to go to fixation.

There is no explanation offered by any of the researchers considered in this book of why about 5% to 10% of women would not have a trait that is seen to be so valuable. Symons used the data on orgasm with intercourse to drive home the variability of the trait, but we do not need to be focused on reproductive sex at all. The high degree of variability within women of the ease and frequency with which they experience orgasm under any circumstances is enough to make the point that female orgasm does not look like a typical adaptation. Some researchers have pursued the idea that the very variability of female orgasm itself may be adaptive. In other words, they have imagined scenarios under which it would not be optimal for women to have orgasm all the time with intercourse (see Chapters 3, 4, and 7). But these explanations still do not apply to women who do not have orgasm at all; nor do they apply to women who always have orgasm with intercourse. In other words, they are like the intermittent reinforcement account of female orgasm, in that the explanation applies only to the subset of women who fall into the "sometimes or often" category of having orgasm with intercourse.

Nothing in this book is intended to deny that there could very well be an adaptive explanation of female orgasm that worked, or that orgasm's contribution to reproductive success may be very cryptic and difficult to ascertain. But as the recently retired director of the Kinsey Institute, John Bancroft, wrote in 1989, there is at present no link between orgasmic potential and fertility in women (1989, p. 86).

I should also emphasize that I am not ruling out the possibility that the chief evolutionary source of female orgasm is to be found in development, but that adjustments of the anatomy or physiology may have occurred. Such alterations are called "secondary adaptations" and it could very well be that selection pressures have shaped details of the female sexual response of orgasm. But at this point, no one has offered such an explanation of female orgasm, so I do not consider it here.

## Standards of Evidence

One of the most common responses to Symons's proposal is to offer an adaptive explanation, and there are several other alternatives proposed by respondents to Symons's account. Hrdy advocates her female-centered theory of female orgasm (reviewed in Chapter 4), while other authors come up with their own untested theories. John Alcock developed one of the theories that came to play an important role in this debate. His account, which rested on females choosing males on the basis of whether they had orgasms with them, was discussed in Chapter 3, where I emphasized the untested nature of its several strong assumptions.

But the key point of interest here is that Alcock is eager to insist that female orgasm serves some *function*. Note that his alternative account seems to concern how women can utilize orgasm, rather than how it came to be present in the population. Such an approach may reflect a small but growing confusion over what is required as prima facie evidence of adaptations. Donald Dewsbury, for example, insists that Symons is inconsistent with regard to what evidence is required to show an adaptation. Dewsbury accuses Symons of adopting an "extremely conservative" interpretation of G. C. Williams's requirements for an adaptation: Dewsbury criticizes Symons for requiring "precision, economy, and efficiency" of an adaptation, emphasizing that such requirements are difficult to apply in practice. "Indeed," Dewsbury argues, "to do so might stifle the study of adaptive significance" (1980, p. 184). Moreover, Dewsbury notes, Symons is not so particular about evidence for adaptation when he is considering other traits, such as susceptibility to jealousy.

Dewsbury's and Alcock's comments are but a preview to a full-blown theoretical debate that occurred after Stephen Jay Gould published, in 1987, an article that advocated Symons's byproduct theory. The next chapter will be concerned with this theoretical debate and its implications for the discussion of the evolution of female orgasm.

# CHAPTER 6

# Warring Approaches to Adaptation

I had been speaking and writing on the case study on evolutionary explanations of female orgasm for two years when I mentioned it in the fall of 1986 to Stephen Jay Gould, with whom I was working on a joint paper on species selection (1993). Gould told me, "Write it up and I'll write a column on it." I went back to the books to add more detail to my analysis, and presented him with a hundred-page critique of the available adaptive explanations (expanded in Chapters 3 and 4 of this book), and a supportive section on Symons, which Gould used to write his controversial column "Freudian Slip," published in *Natural History* in February 1987. In that column Gould strongly supported Symons's byproduct explanation of female orgasm. Gould's entrance into the debate brought the more general issue about when adaptive explanations are feasible to the fore. In this chapter I move to a consideration of these more general issues.

The question, as Gould sees it, concerns the existence of female orgasm. He states that only Symons presents "what I consider the proper answer—that female orgasm is not an adaptation at all" (Gould 1987a, p. 17). Like Symons, Gould uses the male nipple as an instructive analogy. "Male mammals have nipples because females need them—and the embryonic pathway to their development builds precursors in all mammalian fetuses, enlarging the

149

breasts later in females but leaving them small (and without evident function) in males . . . male nipples are an expectation based on pathways of sexual differentiation in mammalian embryology" (1987a, p. 16). He then offers the parallel to female orgasm.

Gould's essay provoked a series of strong responses from John Alcock, which began with a very critical letter to the editor of *Natural History*. The subsequent back and forth between Gould and Alcock brought other commentators into the fray. Most notable among these were Kern Reeve and Paul Sherman, who separately and together tried to help defend Alcock's adaptationist view of female orgasm against Gould's criticisms. In this chapter I review and analyze these debates about female orgasm, which turn on subtly different claims and assumptions both about what an adaptation is, and about what kind of evidence is required to support a claim that a trait is an adaptation. I find that Alcock, Sherman, and Reeve cloud the relevant issues in a host of ways. Most fundamentally, Sherman and Reeve strike an ahistorical posture in their definition of "adaptation," while in practice they resort to the use of a historical definition. I conclude that the debate between Gould and Alcock about the female orgasm thus ultimately depends not on different definitions of adaptation, but on what kind of evidence is appropriate to establishing the existence of a (historical) adaptation. I use this clarification of the debate to show that if Alcock consistently holds to the historical definition of adaptation, Gould clearly has the defensible conception of the relevant evidence, and in turn the most defensible account to date of female orgasm.

## John Alcock versus Stephen Jay Gould

In his letter to the editor, Alcock argued that male nipples and the clitoris are *not* analogous. The difference between male nipples and the clitoris is that male nipples "do not do anything," while the clitoris "does something." He points out that male nipples "do not yield even one percent of the milk of female breasts," the implication being that the male's nipples are but a pale imitation of the female's. This is not the case with orgasms, Alcock argues. "Female orgasm is *not* an imperfect, half-hearted imitation of male orgasm, but a strong physiological response that is different in pattern and timing from male orgasm" (1987, p. 4; his emphasis). Alcock writes, "The clitoris is not an *utterly inert structure;* it is an active participant in a complex and extraordinarily involving event" (1987, p. 4; my emphasis). (Note that Alcock alternates between discussing the female orgasm and the clitoris. This topic will be covered later.) Finally, in a passage that seems to understate the issue dramatically, Alcock writes that "female orgasm is not a guaranteed aspect of sexual intercourse for women, and a certain (modest) amount of cooperation with a partner is generally required for its occurrence" (1987, p. 4).

Alcock then argues that Gould assumes that orgasm must serve the same function in women as in men—that it must serve as a reward for intercourse. As a result, states Alcock, Gould "takes the 'failure' of women to reach orgasm 100 percent of the time as evidence for this *imperfect and nonfunctional nature* of the clitoris" (1987, p. 4; my emphasis). Aside from the fact that the figures don't come anywhere close to 100%, Alcock is right that both Gould and Symons take the low incidence of female orgasm with intercourse as some kind of evidence, but it is evidence only that orgasm is not an adaptation designed to encourage women to engage in intercourse. Gould draws no implications that the clitoris has an "inert" or "imperfect" nature. There is, in fact, nothing in either Gould's piece or Symons's chapter that even suggests that the clitoris is somehow flawed or inert; nor does it somehow follow from the by-product thesis, as I shall discuss in a moment.

As his parting shot, Alcock proposes his mate-choice hypothesis about female orgasm as an adaptive alternative, emphasizing that his view yields testable predictions (as discussed in Chapter 3). He concludes that since the clitoris is not an evolutionary analogue to the male nipple, we should not write off "adaptive possibilities" for explaining the clitoris. "Let's not reject plausible possibilities out of hand," he writes of his own mate-choice hypothesis, against which the developmental theory of the female orgasm needs to be tested (1987, p. 4).

Before proceeding to Gould's response to Alcock, I shall note a few problems with Alcock's discussion. Perhaps most important, Alcock has mistaken the logic of Gould's argument. Gould was not rejecting "plausible" theories for female orgasm in an a priori fashion. Rather, Gould rejects the then-existing adaptationist explanations given the available data at that time, and given the availability of an adequate byproduct account. (Presumably, it is also fair to hold Alcock's suggested alternative to that evidence as well. Recall that he claimed that female orgasm might be an adaptation for female choice of mates. But, as I argued in Chapter 3, his key assumption that orgasm rate goes up with parental investment is unsupported.)

Another problem with Alcock's discussion lies in his understanding of the phenomenon he is trying to explain. He claims that only minor "cooperation" is needed for women to have orgasm with intercourse, but the numbers go against him—again, only at most 54% of women have orgasms more often than not with assisted intercourse, and approximately a third of women never or almost never do. He also claims that female orgasm is "different in pattern and timing from male orgasm"; but, as the masturbation data show, female orgasm itself is not different in timing from male orgasm at all (Kinsey et al. 1953, p. 163).

Finally, what is Alcock doing when he compares male nipples with female breasts? Remember that the major disanalogy between female orgasms and male nipples is supposed to be that male nipples do not "do anything." In addition to the fact that Alcock is neglecting entirely the documented sexual sensitivity and responsiveness of some males' nipples, he relies on a bad analogy. He claims that male nipples "do not yield even one percent of the milk of female breasts" (1987, p. 4). But female breasts are not the same as female nipples. Female nipples do not yield milk at all, the breasts do. What female nipples do is deliver milk; and they can provide pleasure and sexual stimulation for the female during nursing, as well. Since male breasts do not usually experience the necessary hormonal changes, the male nipples do not generally deliver milk—though, in fact, it is well established that they can, under the right circumstances. Male nipples, like female nipples, also can provide pleasure and sexual stimulation for the male.[1] Thus male nipples *can* do what female nipples do, and Alcock's idea that male nipples are a pale imitation of female nipples does not really hold up. Moreover, Alcock's discussion makes it clear that female orgasm is very much like the male nipple. After all, the sensory and excitatory responses are intact in both the male nipple and the female clitoris. In other words, male nipples "do something" in a way very much analogous to the way that clitorises "do something." But none of the adaptationists, including Alcock, attempts to insist on an adaptationist explanation for male nipples.

Let us proceed to Gould's response to Alcock. Gould focuses on Alcock's objection that the clitoris is neither "inert" nor "imperfect and nonfunctional" in nature. At first blush, it seems odd that Alcock is attributing a claim about imperfection and nonfunctionality to Gould at all. There is nothing in the byproduct account per se that would yield the conclusion that the clitoris is in any sense imperfect; there is only the conclusion that it would be expected to perform in certain ways under certain conditions. Gould tries to make sense of this inference by pointing out that Alcock's notion of perfection is concerned with the *current function* of orgasm, rather than with its historical origin. Gould writes, "This false inference exemplifies my major complaint about adaptationism—its *logically* incorrect equation of current utility with reasons for historical origin" (1987b, p. 4). There are many more traits that are currently useful than there are traits that are directly connected to reproductive success. Adaptationists, according to Gould, think of the two categories as coextensive, and focus on "only those structures that natural selection builds or maintains for current function" (1987b, p. 6). Note here that Gould is using the term "function" in a technical sense. A function is an action performed by a part of an organism, where the part was directly selected in the past to do that particular action. Thus a fish's gills have the function of extracting oxygen from water in this sense, because they were directly selected in the past to perform this action.[2]

The problem with focusing only on traits that are naturally selected for their current function is that traits that enable an organism to perform actions that are useful but not naturally selected for get lost or miscategorized in the shuffle of doing science.

Note that Gould does not equate current contributions to fitness with "current utility," which is much more inclusive, covering some traits that do not contribute to current fitness. (I shall use the terms "fitness" and "reproductive success" interchangeably.) Gould gives several examples of such useful but nonselected traits, including the human behavioral traits of reading and writing. As he points out, "Natural selection did not act specifically for these foci of technological societies," even though these traits are clearly of vital importance today in some cultures—even though they have "current utility." Gould argues that because Alcock equates "origin as an adaptation" with "current utility," he gets confused about the byproduct theorists' claim about female orgasm. As a result, Alcock thinks that when Gould says the female orgasm is probably nonadaptive Gould is saying that the clitoris itself is currently useless— or, in Alcock's words, "imperfect and nonfunctional." (Alcock also slips into talking about the clitoris, a mistake that has consequences for his arguments.) But Gould explicitly acknowledges that the clitoris has "vital utility," especially in causing sexual excitement as well as causing the female to have orgasms. In other words, a trait's current utility simply does not imply that it has an evolutionary function. Conventionally, evolutionary functions require that there were past selection pressures that built the trait over evolutionary time. Currently useful traits include both those with evolutionary functions and those without. Thus we get Gould's and Symons's view that female orgasm is currently useful but probably does not have an evolutionary function.

It is crucial to note that at the heart of Gould and Alcock's disagreement about the female orgasm is a deeper question about whether current utility by itself is always strong evidence of evolutionary function. According to Gould, the adaptationist bias of focusing all evolutionary interest on those structures for which one can make a case that they have been built for current evolutionary function must be based on the prior assumption that current utility and evolutionary function always come together. Hence, any other structures that may be currently useful but that are not built or maintained for current evolutionary function are either wedged into incorrect adaptationist explanations or cast out of the realm of interesting traits precisely *because* they do not serve current evolutionary functions. In contrast, Gould believes that current utility is often separate from evolutionary function: "Adaptations are features built by natural selection that enhance reproductive success; the domain of biologically useful structures is vastly greater" (1987b, p. 6).

I will return to the question of whether Alcock and Gould are really disagreeing in their definitions of "adaptation." For now, note that Alcock and Gould seem to be disagreeing about the answer to an empirical question: whether there are many structures that were not evolved by natural selection to perform their present activities. This issue has profound implications for the appropriate approach to researching evolutionary questions. According to Alcock, the very fact that the clitoris does something (has current utility), participates in causing orgasm on some occasions, implies that it was selected to do what it does. His assumption is that natural selection brought the orgasmic structures to their present state; the problem is to figure out which selective hypothesis is correct. On Gould's approach, the fact that the clitoris does something is adequately accounted for by the claim that the penis was and is selected to do something. Under the byproduct account, we would not expect the clitoris and associated structures to do nothing; on the contrary, we would expect them to have the potential to perform an action homologous to that which the penis and associated structures do. Alcock appears to accept the fact that the clitoris and penis are homologous organs; nevertheless, because the female orgasm is frequently used, it cries out to him for an adaptive account.

"The adaptationist position is an invitation to scientific investigation," writes Alcock, and this is meant to contrast with the byproduct account, which, it is implied, stifles inquiry prematurely, resting with an unsatisfactory, incomplete explanation (1987, p. 6). This is Alcock's reasoning, then: a developmental explanation may account for many structural similarities, but the fact that female orgasm plays a role in human lives demands that explanations of its role in reproductive success be investigated.

### Further Debate: Paul Sherman and Kern Reeve

The core disagreement lies in whether a byproduct account can be considered an appropriate evolutionary explanation of a trait, or whether such an account automatically ought to be supplemented with an adaptive account if the trait has any significant current utility. This topic has been the locus of a great deal of discussion over the last few decades. The female orgasm example serves as an excellent case study of the various theoretical issues involved in these debates. In the rest of this section, I shall discuss several of these theoretical issues, in a dual effort to shed light on the larger controversy

and to make progress toward resolving the question of why female human beings can have orgasms. The central topics include
the various definitions of adaptation in use; the standards of evidence for demonstrating that a trait is an adaptation; and the assumptions made regarding fitness differences. A number of authors
have weighed in on this debate between Gould and Alcock, and several more have written on the topics directly at stake between them.
In this section I shall highlight some of these analyses.

The first question to be broached is whether Gould and Alcock
are using different definitions of "adaptation." It turns out that
there are a number of subtly different meanings of the term in use,
as discussed by Burian (1983; 1992), West-Eberhard (1992), and
Brandon (1990), among others. There is one definition of what we
could call "engineering" adaptation, which is shared by Darwin,
George C. Williams, Gould, Richard Lewontin, Richard Dawkins,
and Symons (Lloyd 2001). Under this definition, "it is correct to
consider a character an 'adaptation' for a particular task only if
there is some evidence that it has evolved (been modified during its
evolutionary history) in specific ways to make it more effective in
the performance of that task, and that the change has occurred due
to the increased fitness that results" (West-Eberhard 1992, p. 13).
She adds, "Incidental ability to perform a task effectively is not sufficient; nor is mere existence of a good fit between organism and environment. . . . To be considered an adaptation a trait must be
shown to be a consequence of selection for that trait" (1992, p. 13).

West-Eberhard then offers a nice review of the types of evidence
that may be given to determine whether a trait is an adaptation.
Clearly, under her definition—which is similar in spirit to Burian's
and Symons's[3]—evidence that a current trait is an adaptation requires a historical account involving the trait's past participation in
an evolution-by-selection process. This historical approach to adaptation should be contrasted with the approach taken, for example, by Clutton-Brock and Harvey, who define an adaptation "as a

difference between two phenotypic traits (or complexes of traits) which increases the inclusive fitness . . . of its carrier" (1979, p. 548; cited in Burian 1992, p. 10). On this latter definition, historical evidence appears to be unnecessary to demonstrate that a trait is an adaptation. One only needs to show that the trait *presently* contributes to an organism's reproductive inclusive fitness. I shall call these two approaches the "historical" and "current-fitness" approaches, respectively. Note that the historical approach also includes claims about the current-fitness consequences of a trait: That is, according to West-Eberhard, having a current contribution to fitness is one of the requirements that an adaptation must fill. It's just that the historical approach also requires evidence about past contributions to fitness.

On the face of it, Gould and Alcock might be thought to be using incompatible definitions of adaptation and thus talking past each other: Alcock emphasizes a current contribution to fitness, while Gould focuses on the historical origin of female orgasm. This seems to be Paul Sherman's view of the Alcock-Gould debate. Sherman, a leading sociobiologist, thinks Alcock is saying that clitorises and orgasms are adaptations if "they seem designed to enhance reproductive success," whereas Gould's favored hypothesis is silent regarding the relationship between the clitoris and female fitness (Sherman 1988, p. 616). As we shall see, Sherman disagrees with Gould about two issues. One is whether Gould and Alcock are using compatible definitions of adaptation. According to Sherman, Gould makes the mistake of pitting the developmental byproduct account of female orgasm against a claim based on current differential reproduction. Sherman is thus claiming that Gould and Alcock are using incompatible definitions of adaptation, and should thus be expected to disagree. The other disagreement between Gould and Sherman concerns the status of the search for a connection between orgasm and reproductive success, which I will discuss shortly.

Interestingly, Sherman chides Gould for refusing to acknowledge that the byproduct hypothesis "predicts that the clitoris is essentially neutral for female reproduction: a testable and falsifiable prediction" (1988, p. 618). Note that Sherman is discussing the fitness consequences of the *clitoris* and not the orgasm. This is a significant confusion, because no one is arguing that the clitoris—in its role of producing sexual excitement in the female, thereby promoting her to engage in sexual activity—does not play an important role in female fitness. In fact, Gould and Symons explicitly claim just that role for the clitoris, so Sherman is actually attributing a view to them that they explicitly deny. Still, substituting orgasm for clitoris, Gould and Symons both considered the evidence—or lack of it—regarding any potential correlation between female orgasm and reproductive success. Not only is the prediction that orgasm is neutral with relation to reproduction "testable and falsifiable," but it is in accord with the evidence. Thus it is hard to see what Gould is "refusing" to acknowledge.

Sherman then offers his own reformulation of Niko Tinbergen's four crucial research questions in the field of animal behavior in order to make the claim that Gould and Alcock are really asking different questions. According to Tinbergen, the field of animal behavior is shaped by four distinct but overlapping questions. They include questions of "causation," or the physiological pathways of behavior; "ontogeny," the study of the "change of behaviour machinery during development" (Tinbergen 1963, p. 424); "survival value," the study of the function and effects of behavior; and "evolution," the study of the course and dynamics of evolution (1963).

According to Sherman, Tinbergen's question concerning the "evolution" of a trait is properly interpreted as a question regarding the evolutionary origins of that trait, a plausible enough suggestion. But rather than letting Tinbergen's "survival value" question involve just a contemporary physiological analysis of the function of the parts or behaviors of an organism, Sherman describes the question as one of functional consequences, and interprets it as an issue concerning current-fitness consequences of a trait. He then claims that both the evolutionary origins question and the functional (fitness) consequences question are "ultimate" questions in Ernst Mayr's sense: they concern how a trait is causally involved in the evolution of an organism. This agrees, to some extent, with Tinbergen, who emphasizes that evolution is studied partly by studying the survival value of species-specific characters. However, Tinbergen emphasizes the difference between establishing a selection pressure responsible for maintaining a trait, and establishing a selection pressure responsible for molding the trait in the first place, noting that one cannot simply infer from one to the other (1963, p. 429). Tinbergen also emphasizes the connection between studying survival value and studying evolution: he notes that the study of survival value is important because of the need to know how to judge fitness, which is in turn needed in order to understand the part played by natural selection in evolution of the trait (1963, p. 427).

On Sherman's view, the evolutionary history of a trait is a different "level of analysis" from the functional consequences of a trait: Gould and Alcock are asking and answering Tinbergen-like questions at different levels of analysis, the answers to which may not be mutually exclusive. Thus, according to Sherman, their views are compatible; Gould's explanation was directed at the evolutionary origin of the clitoris (orgasm) whereas Alcock's hypothesis—the one about females using orgasm for mate choice—concerned present functional consequences.

Alcock, interestingly, gives an analysis different from Sherman's in describing how he and Gould arrive at compatible answers to different questions. Alcock emphasizes the difference between "proximate," immediate developmental or physiological origins in an individual, and "ultimate" questions concerning the adaptive or reproductive value of the traits (1998, p. 328). Alcock thinks Gould is arguing that "because the adaptationist is interested in evolutionary or ultimate explanations, he or she would be out of business if it could be shown that a proximate explanation of a trait makes it unnecessary to explain why selection resulted in the spread of the mechanisms underlying the trait" (1998, p. 328). In other words, by confusing the complementary roles of proximate and ultimate explanations, Gould is making a fundamental error in biological reasoning.

Note what is required for Alcock to hold this view. He must believe that the byproduct view is not an evolutionary view at all, despite the fact that it offers a historical account for the presence of the trait in the population. He seems to think that Gould is giving only an ontogenetic story, an account of how human females develop orgasms through their development from embryo to adult. Alcock is attempting to rule the byproduct explanation out of court as a competing explanation to the adaptationist view that he supports, by misinterpreting the basic claim of the byproduct view. Consider the following: "proximate explanations of a biological characteristic do not make it impossible to ask whether the trait of interest contributed to individual reproductive success in the past or does so currently. If we were to discover the female orgasm occurred with positive effects on female reproductive success, we would gain an *evolutionary* dimension to our understanding of this trait that is not covered by *any* proximate explanation" (1998, p. 330; my emphasis; his emphasis). Note here that Alcock seems to require that the trait be correlated with reproductive success in order for an explanation to be "evolutionary." In other words, he is insisting that all evolutionary explanations be adaptive explanations. He cannot see that a byproduct explanation *is* evolutionary; thus he eliminates it as an alternative to an adaptationist account.

Alcock's view is much more extreme than Sherman's. Sherman is arguing that the byproduct account does have to do with the evolution of the trait of female orgasm, but not with its current contribution to fitness. In what follows, I shall focus on this more plausible view of Sherman's.

In a further intervention, Ian Jamieson challenges Sherman's reworking of Tinbergen's questions, and disputes the claim that Alcock and Gould have compatible theories. As Jamieson argues, it is reasonable to see Alcock and Gould as offering competing selectionist hypotheses "concerning the question of how the clitoris and orgasm evolved" (1989, p. 696). On Gould's favored account, the original selection pressure was directed at the penis and not the clitoris, and while the clitoris plays an "intricate role in facilitating sexual pleasure in women," it probably does not contribute to differential reproductive success in the female (Jamieson 1989, p. 696). Note that Jamieson, too, is confused about the actual trait under contention, the ability for the clitoris and other tissues to perform the reflex behavior of orgasm, rather than just provide sexual pleasure. Again, no one has argued that the clitoris—as the primary source of sexual pleasure in females—does not contribute to reproductive success.

Nevertheless, suitably corrected, Jamieson contrasts usefully the byproduct account with Alcock's own selective account of female orgasm. Jamieson reads Gould's and Alcock's accounts as two mutually exclusive hypotheses about the evolution of female orgasm, one in which the orgasm is a result of the indirect selection on the male, and one in which female orgasm was itself directly selected and "is designed to enhance reproductive success" (1989, p. 696).

Sherman rebuts Jamieson's analysis, claiming that Gould was concerned with evolutionary origin while Alcock dealt with "the likely effects of clitorises and orgasms on female fitness" (1989, p. 697). According to Sherman, these hypotheses cannot be compared, and Gould confused the issue by acting as if they could. Gould's argument, writes Sherman, concerned how orgasms originated, and concluded that "hypotheses about how clitorises [*sic*] affect the fitness of human females are superfluous" (1989, p. 697). Once we replace "clitorises" with "female orgasm," this is partially correct; Gould's implication—and Symons's—is that once an adequate developmental explanation for the appearance and persistence of the female orgasm is given, further selectionist explanations are unnecessary, because there is no known correlation between female orgasm and reproductive success.

Alcock and Sherman repeatedly assert that under the byproduct account the female orgasm would deteriorate because it was not under direct selection. But this view is mistaken. It is vital to understanding the byproduct theory to see that it accounts for both the appearance and the persistence of the trait of female orgasm. Given the assumption of constant selection pressure on males for the equipment underlying male orgasmic capacity, one would expect a resulting constancy of female orgasmic capacity. Alcock and Sherman are forgetting that under the byproduct hypothesis, there is, at minimum, steady selection on orgasmic capacity in males that directly affects the embryological states relevant to female orgasmic capacity. This issue of maintenance is one of their key arguments against the byproduct approach, and it is profoundly confused.

At any rate, Sherman argues that the existence of the embryological account given by Gould does not mean that questions cannot still be asked about the functional design, persistence, and reproductive consequences of female orgasm. Of course, this is correct. Questions can be asked. But at the time that Sherman was writing, neither Alcock nor Sherman had given us reason to think that such questions might be fruitful, given that there was no evidence linking female orgasm with reproductive success. Moreover, Alcock and Sherman both seem to think, incorrectly, that such evidence had not even been weighed by Symons or Gould. But we have many decades of sex research, most into reproduction, fertility, and their ties to sexuality, which failed to produce any evidence linking orgasm with fertility. This lack of connection is well established among sex researchers (for example, Bancroft 1989; but see Fox et al. 1970 and Singer 1973 in Chapter 7 below).

Alcock and Sherman accept the embryological account, but seem to want to add an additional adaptationist account. Such a possibility is not ruled out by Gould, who acknowledged that a "secondary adaptation" could have occurred, though there is as yet no evidence to support such a claim. Oddly, the notion of secondary adaptation—refinement of the detailed features of female orgasm and its occurrence—has not been actively pursued by anyone, although it is an important aspect of the conceptual space in this case.

One ill-considered point is Sherman's claim that even if most female primates have clitorises and orgasms (not the claim under discussion, but I'll ignore that), hypotheses about fitness effects are still relevant, because orgasms could result from consistency of selected function instead of phylogenetic inertia (1989, p. 698).[4] We have already seen the problem with this claim; consistency of selection pressure on the *males* is presumed to produce whatever consistency of performance there is among the females—Gould's and Symons's argument does not concern phylogenetic inertia at all. It is actually a developmental-selectionist account, because of its reliance on the selective pressure on the male. Moreover, Symons and Gould considered many claims supposedly regarding function in the context of various proposed explanations of female orgasm, all of which were faulty. This failure does not mean that there *could* be no evidence supporting an adaptive account of female orgasm; such a strong claim is not needed here. It suffices that there is already an account of the maintenance of the trait contained within the developmental-selectionist account. Sherman fails to establish that an additional, adaptive account is either necessary or suggested by any evidence. Despite this failure, Sherman concludes that Gould's "summary dismissal" of all adaptationist hypotheses for female orgasm was "inappropriate."

Part of what is going on here is a struggle both over scientific authority and over rival conceptions of proper scientific methodology as well as over which scientific explanations to pursue. Sherman and Alcock are accusing Gould of prohibiting scientific inquiry by resting on an inferior, byproduct explanation; conversely, Gould views Sherman and Alcock's inquiry into adaptations as unnecessary given the available evidence for a developmental-selectionist explanation of the trait.[5] For Gould, the question must be: Is there enough evidence tying the existence of orgasm to reproductive success to warrant pursuit of the kind of hypotheses favored by Sherman and Alcock? In 1987, Gould considered the available evidence and did not consider it sufficient to pursue these hypotheses. The situation now is altered by the work of Baker and Bellis and Thornhill and colleagues, whose data I shall consider in the next chapter. However, Gould's analysis in 1987 clearly reflects the information provided by his sources (1987a, pp. 16–18). Gould rejected the available adaptive hypotheses for good reasons, and not summarily.

Part of Sherman and Alcock's position is driven by an intuition not shared by Gould or Symons. Sherman writes that "Alcock (1987) had good reason to question how a structure that plays such an intricate role [in facilitating sexual pleasure in women—here, he is talking about the clitoris again] and one *so obviously related to fitness* could possibly be reproductively neutral" (1989, p. 698; my emphasis). Again, no one has claimed that the clitoris is reproductively neutral. Focusing on the orgasm, though, does not help matters, since the adaptive stories mentioned and criticized by Gould, Symons, and me attempt to connect orgasm with evolutionary fitness. Moreover, there is the overwhelming lack of evidence for a connection between fertility and orgasm from the sex research literature. The crucial point is that, whatever people's intuitions are regarding the obviousness of what should be the case in evolution, I have already argued, in relation to the adaptive stories current at the time, that there is no credible evidence that orgasm correlates with reproductive success. Perhaps intuitively, all of the details of sexual performance should have something to do with reproductive success. Symons, however, presented arguments that orgasm in the female does not. The fact that such arguments have been appealed to in the formulation of Symons's byproduct hypothesis cannot legitimately be ignored simply because they do not accord with either Sherman's or Alcock's intuitions.

Sherman's argument is that Gould's and Alcock's claims require different "levels of analysis," supposedly because Gould is making a claim about evolutionary history while Alcock is concerned with current-fitness consequences. The real question, then, is whether Alcock or Sherman thinks that current-fitness consequences have *implications* for the history of a trait. Let us consider, then, the uses to which current-fitness evidence can be put.

### Current Fitness

Take Alcock's appeal to current-fitness evidence. Is he simply trying to claim that we should examine current fitness to see whether we have an adaptation? This might imply that he holds a typical historical definition of adaptation, under which evidence of current-fitness contribution is one of two requirements. Or is it that current-fitness evidence can itself be used to support a historical selection account? The way this would work is that the evidence from current environments about relative fitness of traits would be used to support claims about the past; if past environments and selection pressures were similar to present environments and selection pressures, then we could use current-fitness data as evidence for an account of past selection. Take an example: The current-fitness consequence of mammalian fur is that it helps with thermoregulation, and we can observe and experiment on this in current environments of varying temperature. One might infer that some past mammalian environments contained varying temperatures, and these created a selection pressure for better thermoregulation, which, over evolutionary time, led to the current state of fur in a species. Thus current-fitness consequences *may* be used to suggest past selective regimes, and thus they can contribute to a historical account of adaptation.

There are always risks and problems associated with using current fitness as evidence for past adaptive scenarios, however. Most fundamentally, we are, no doubt, sometimes mistaken in the assumption that past environments for the organism were like present ones. In other words, the *uncritical* assumption of the constancy of selection pressures can be problematic. Without additional independent historical evidence—perhaps from paleobotany or geology—there is no telling whether past environments held the same challenges, during the relevant evolutionary period, as they do now. Moreover, the human case is especially difficult: the last 10,000 years have seen enormous changes in numerous aspects of our environments and how we interact with them (Symons 1990). Nevertheless, this recent span containing dramatic changes for most of our species covers but a fraction of the time that *Homo sapiens* has evolved. Also, from what is known now, our species seems to have reached its current physical state hundreds of thousands of years before the period of technological development began. Thus inferences from our present place in nature to what our place in nature may have been like during relevant periods of our evolutionary history can be tendentious, at best, without appropriate research.

There are other problems with using current reproductive success as an indicator of adaptation. The usual way of assessing the current-fitness contribution of a trait is to correlate variation in the trait with differences in reproductive success. Symons gives a long list of reasons for rejecting this methodology as a step in inferring about past adaptation (1990, p. 430). These include: that one could have correlations that are too small to detect, but that have big selective significance over evolutionary time; or one could have spurious correlation due to a common cause. In addition, "a given trait may promote fitness—and hence correlate positively with reproductive success—because it currently produces some effect other than its evolved function" (Symons 1990, p. 430). This would be what Gould and Elisabeth Vrba (1982) dubbed an "exaptation." Symons also adds, the hypothesis that a trait is an adaptation does not imply that the trait is currently adaptive.

This is a peculiar thing to say, on the face of it, and it runs contrary to definitions like West-Eberhard's that do require some current contribution to fitness (in addition to historical contributions) in order for something to count as an adaptation. Under the usual historical definitions, Symons is calling something that *used* to be an adaptation, an adaptation. This is not so unusual in practice, but it should be made explicit. After all, in Hrdy's defense of her multiple-male theory of female orgasm, she emphasizes that it is a *past* adaptation for a mating system no longer in place. Speaking precisely, such adaptations should be called "past adaptations," in order to distinguish them from traits that currently contribute to survival and reproduction. Note that the concept of something that used to be an adaptation requires use of the full-fledged historical definition of adaptation. In sum, Symons argues that correlational studies between current trait values and current reproductive success by themselves do not provide "direct evidence" for adaptation (1990, p. 431; Dewsbury 1992, p. 103). This view seems in conflict with the view of Thornhill, who argues that "the evolutionary purpose/function of an adaptation can be studied productively without any reference to or understanding of the adaptation's origin" (1990, p. 32). This is done by studying the trait's "functional design," a study that Thornhill calls "teleonomy." I assume that Thornhill means the same thing as Symons's "engineering analysis," in which "design is recognized in the precision, economy, efficiency, complexity and constancy with which effects are achieved"

(Symons 1990, p. 429). Thornhill concludes that "teleonomic analysis provides *direct evidence* of how long-term evolution works because adaptations contain actual information about long-term evolution" (1990, p. 45; my emphasis). But functional analyses do not provide "direct evidence" about a historical past, despite Thornhill's claim that the study of "true functional design" "demonstrates how the trait covaried with fitness in the environment of evolutionary adaptation" (1990, p. 41). Symons recognizes that collateral evidence for the assumptions about environment and selection pressure is necessary. Still, Thornhill agrees with Symons that one cannot use current selection to show the evolutionary history of a trait; he cites Grafen (1988) as giving a decisive critique of analyzing adaptation "by studying individual variation in reproduction" (1990, p. 49).

Reeve and Sherman take an entirely different approach to evidence about current fitness. They present an ahistorical definition of adaptation: "An adaptation is a phenotypic variant that results in the highest fitness among a specified set of variants in a given environment" (1993, p. 1; see Wade 1987). They complain that historical definitions of adaptation do not include traits that are maintained by natural selection, nor do they answer the question: "Why do certain phenotypes predominate in nature" (1993, p. 1). But it turns out that their approach to adaptation is not ahistorical, as advertised. They use current fitnesses to explain why certain traits predominate over others conceivable in nature, "and then *infer evolutionary causation based on current utility*" (1993, p. 2; my emphasis). In other words, in direct contrast to Symons, Williams, Gould, and West-Eberhard, Reeve and Sherman want to use current fitness to infer not just present adaptation, but adaptive history. But the advantage of their approach, they claim, is that it "decouples adaptations from the evolutionary mechanisms that generate them" (1993, p. 1). In other words, their definition licenses, they think, the study of present adaptations in isolation from the trait's history. Nevertheless, they also want to use current fitness to infer the evolutionary history. Thus their view suffers from an internal tension between these two views of the role of current fitness effects.

Reeve and Sherman use their analysis to attack Gould's views on female orgasm. They accuse him of calling female orgasm a "non-adaptation" rather than what they think it really is: an "exaptation." Let me take a moment to review the concept of exaptation, which was introduced by Gould and Vrba in 1982 in order to counteract the oversight or misrepresentation of some traits by biologists. According to Gould and Vrba's terminology, traits that currently contribute to reproductive success are "aptations"; aptations come in two kinds, adaptations and exaptations. Adaptations were directly selected and evolved for their current function, while exaptations were not, but are nevertheless contributing to reproductive success. The term *exaptation* itself applies to two types of traits: traits that were adapted for one evolutionary function, but were later coopted to serve another function; and traits that were correlates of growth or accidental byproducts that were later coopted to serve a role in current reproductive success. In both cases, the trait is understood as currently contributing to fitness.

Reeve and Sherman's accusation of a mistake is premised on the assumption that orgasm contributes to current fitness, a claim that Gould never acceded to, because of orgasm's apparent lack of connection with current fitness. The accusations also seem to misunderstand Gould and Vrba's categories, which do not mesh well with Reeve and Sherman's own, newly suggested ones. On Gould and Vrba's view, exaptations and adaptations are distinct, nonoverlapping categories:[6] exaptations may later turn into adaptations for their current function if there is selection pressure to modify the trait, but exaptations are, by definition, not adaptations. This is why Gould calls an exaptation a "nonadaptation." The fact that Reeve and Sherman propose to use "adaptation" (where Gould and

Vrba use "aptation") to include both adaptations and exaptations does not make Gould mistaken in his categorization. Reeve and Sherman conclude that the historical definition of adaptation is too conservative, and that "the concept of adaptation may usefully be broadened to include traits for which there is no demonstrable history of selective modification" (1993, p. 7). Thus it is clear that Reeve and Sherman are suggesting a programmatic change in the use of the term "adaptation" and in the treatment of evidence for adaptation (Rose and Lauder 1996; Sinervo and Basolo 1996). It also seems clear from their paper that Gould's treatment of the female orgasm case is a motivating factor for such a terminological shift. Their suggested change in the meaning of "adaptation" has not been accepted by many evolutionary biologists outside the field of animal behavior. Neither Gould nor Alcock seems to accept the definition, so it turns out to be a red herring.

Where does all this leave us? There appear to be at least two major definitions of adaptation in use: an ahistorical one and a historical one. Although Symons's and Gould's opponents advertise themselves as employing an ahistorical definition of adaptation, their own proffered standards of evidence reveal that in practice they hold a historical conception of adaptations as well. The disagreement thus appears not to be over what counts as an adaptation, but rather over what evidence may be used to establish that a trait is an adaptation, in the historical sense. But when we examine the details, it seems that the biggest differences lie in how the various authors want to treat evidence of current-fitness contributions. Some, like Symons, want to treat current-fitness evidence as only weakly relevant; he focuses on engineering analyses and comparative evidence. Others, such as Alcock and Sherman, argue that *because* clitorises and orgasms currently contribute to female reproductive success, they are the probable outcome of past natural selection (1994, p. 59). In other words, they are using (putative) current fitness alone *as direct evidence* for past evolutionary processes (see

Alcock 1998 and Sherman 1989). Both sides agree, at least implicitly, that some historical account needs to be given; the differences lie in what the best (or even adequate) methods are for producing and supporting such an account. Thus I agree with Jamieson (1989) and Mitchell (1992) that Alcock and Gould's debate concerns genuinely conflicting accounts of the historical explanation of female orgasm. Contrary to first impressions, Alcock does not rely ultimately on current fitness alone, but means to use it as evidence for a consistent historical selection pressure, as is clear from his later discussion (Alcock and Sherman 1994, p. 59).[7] Reeve and Sherman's attempted reworking of the terminology of adaptation serves as a red herring in the debate between Gould and Alcock, as we have seen. The real difference between the parties to these debates lies in the sort of relation they would like to draw between current-fitness evidence and evolutionary history, and does not, as Sherman, Reeve, and Alcock think, centrally concern different definitions of adaptation. Thus the first conclusion we ought to draw is that so long as we are ultimately after an historical account of the existence of female orgasm, Gould and Symons are right to insist that adaptive accounts provide historically relevant evidence.

There is another factor at play in the debates, though, and it concerns the assumptions that the contestants make about the status of evidence linking female orgasm in particular with reproductive success. Some authors assume there actually *is* currently good evidence of a connection between female orgasm and fitness differences, while others assume there is little or none; this causes some confusion about where the burden of proof lies.

Symons, Gould, and I each approach the byproduct hypothesis for female orgasm by documenting that not only is there no good evidence linking female orgasm with current fitness consequences, but there is much evidence already available from sexology and from the nonhuman primates suggesting that there is no such connection. Several of the interlocutors in these debates simply assume the opposite, however, and this makes Symons's position look arbitrary to them. For example, Alcock and Sherman describe their own thesis as: "Clitorises and orgasms *currently contribute* to female reproductive success as the possible outcome of past episodes of natural selection" (1994, p. 59; my emphasis). Such a position accords with the adaptationist approach advocated by John Tooby and Irven DeVore, who claim that "one begins with the methodological presumption that the great majority of significant traits are or were *adaptive,*" and the investigator attempts "to trace the adaptive consequences of a feature" (1987, p. 194; my emphasis).

Douglas Armstrong states that Sherman's position is that clitoral orgasm "could improve inclusive fitness, and such functional consequences could explain the evolution of [this trait] by means of natural selection" (1991, p. 824). Armstrong emphasizes that "Smith (1984) discusses the functions of reproductive structures to speculate on how selection in the context of potential or actual sperm competition may have operated in evolutionary history," arguing that "leading researchers" are interested in using information on the function of the clitoris to explain the evolution of it. Thus Armstrong seems to be optimistic about the promise of investigating current-fitness consequences of the clitoris (but let's say female orgasm) to explain its evolution, based on the past successes of a research program.

In a related vein, Donald Dewsbury claims that Gould goes too far in supporting a nonadaptive account of female orgasm, arguing that "we need to study the consequences of orgasm for differential reproductive success and *then* determine whether a plausible case can be made for drawing the loop from present consequences to the past history of natural selection. These need to be studied, *not asserted or denied a priori*" (1992, p. 103; my emphasis). Andrews and colleagues extend this point, emphasizing that an adaptationist "testing strategy" is necessary for the byproduct view, since "demonstration that the female clitoris and orgasm are byproducts requires the failure to find evidence for its special design" (2002, p. 32). But Alcock believes something stronger, namely, that evidence showing a connection between female orgasm and reproductive success is already in. He writes, "The fact that female orgasm apparently can draw sperm closer to the cervix, and thus potentially affect the fertilization chances of an egg (Baker and Bellis 1993[b]) is evidence of the kind of complex functional design that demands adaptationist analysis" (1998, p. 330). Alcock also cites Hrdy (1988) and Thornhill et al. (1995), noting that these researchers "have not let Gould stop them from developing and testing alternative hypotheses on the possible current utility of female orgasm" (1998, p. 334). Thus Alcock, Sherman, Armstrong, and Dewsbury all assume either that there is or is likely to be evidence linking orgasm to reproductive success.

Once again, we have a conflict of assumptions about the status of the evidence. Symons and Gould take it that the kind of evidence reviewed in the previous chapter tells against female orgasm's making a difference to reproductive success, while the critics' position is rather that no such evidence has been studied, or that it has been, and somehow tells in favor of their own view.

Note that even early on, when there was no evidence of any association between female orgasm and current fitness, Dewsbury and Armstrong both remained confident not only that such evidence was likely in the future, but that no evidence for or against the association had been considered by Symons or Gould. These expectations on Dewsbury's and Armstrong's part can in themselves be useful in understanding these authors' opposition to the Symons and Gould views. If one is optimistic that such evidence will be forthcoming, then Gould and Symons appear dogmatic and unscientific. This impression is made worse if one ignores the evidence that Symons gave in support of his claim that female orgasm and current fitness seem to be unrelated.

The situation is rather different for the later papers, particularly for Alcock's; Alcock cites research done in the mid-1990s by Baker and Bellis and Thornhill and colleagues, which claims to demonstrate actual connections between female orgasm and reproductive success. Whether or not such evidence is enough to infer a selective history for female orgasm remains an open question under these findings, because of the aforementioned problems with inferring past evolution from current-fitness differences. Nevertheless, such findings change the shape of the debate, if in fact they document genuine fitness consequences for female orgasm. The next chapter will therefore be devoted to a careful examination of these claims.

## Adaptationist and Byproduct View Supporters

At this point, it is useful to distinguish various types of commitment to finding adaptive explanations. One group, which I call the "cavalier" adaptationists, and which includes Morris, Beach, and possibly Sherman, and the authors in Chapters 3 and 4, starts with the assumption that the trait of female orgasm is an adaptation. As such, the cavalier adaptationists are pursuing an adaptationist methodology, in which inquiry into a trait starts with the assumption that the trait under investigation *is* an adaptation (Mayr 1983). The problem with the cavalier adaptationists is that they do not inquire further into the trait, in order to establish that it really is associated with reproductive success. This is in contrast to those we could call the "conservative" adaptationists, who start investigating a trait by assuming that it is an adaptation, but who then pursue evidence that the trait really does contribute to reproductive success (a method advocated by Tooby and DeVore 1987; Sinervo and Basolo 1996). We have few examples of such researchers investigating the trait of female orgasm, for example, Symons and Gould. A third group, which I call "ardent" adaptationists, after Alcock's self-title, have a more restrictive approach to research. Ardent adaptationists avidly pursue adaptive accounts of a trait and, in fact, will not be satisfied with an account of female orgasm that is not adaptive. Ardent adaptationists would define themselves as enthusiastically pursuing adaptive explanations, but would not declare that no account will do unless it is adaptationist. Nevertheless, this is what their view on female orgasm actually amounts to, as expressed in their own analyses. They do allow that other traits, such as the male nipple, are byproducts of development and not adaptations. Alcock, and probably Reeve and Sherman, are ardent adaptationists (Alcock 2001). The dogmatic nature of ardent adaptationism about female orgasm is discussed in Chapter 8.

As we have seen, ardent adaptationists and byproduct view supporters treat the lack of evidence about the correlation between female orgasm and reproductive success differently. The ardent adaptationists think either that there is positive evidence about such a correlation or that positive evidence is bound to be forthcoming. The byproduct view supporters pay attention to the fact that over 80 years of sexology research and nearly 40 years of evolutionary research have failed to produce such a correlation, *pace* the sperm-competition data. In this case the burden of proof is on the adaptationists, who are obliged to come up with evidence linking reproductive success with female orgasm.

The ardent adaptationists and byproduct view supporters also have different perspectives on the supportive evidence for the byproduct view. As I emphasized in the previous chapter, the byproduct account leads to expectations that have been fulfilled. Nevertheless, the adaptationists continue to behave as if there is no supporting evidence for the byproduct view, seeing the view as being essentially negative: they see the byproduct view as claiming only that no adaptive account has been found, rather than as a positive story in its own right. Hence their attitude that adopting the byproduct view is tantamount to "giving up" on an evolutionary account altogether. One way of handling the evidence supporting the byproduct view is to behave as if it is evidence only for a proximate account of how female infants grow up to have orgasms, when they do, as we have seen Alcock do. Alternatively, Sherman takes the positive evidence and allows that it may support a historical evolutionary story, but claims it does not count as addressing the primary question of the adaptive function of female orgasm. This is where their attitudes about the evidence for a correlation between reproductive success and female orgasm really have some bite. These complaints about the byproduct view's not taking into account the adaptive function of female orgasm simply assume that such an adaptive account is called for by the evidence, an assumption that is seen as highly problematic from the point of view of the byproduct account, or from point of view of the four requirements for an adaptation outlined in Chapter 1.

There is a central question here, which is: Under what conditions would the ardent adaptationists accept a byproduct account of female orgasm? It seems clear from their writings that no such account will ever be acceptable. Part of that position is supported by their idiosyncratic views on the acceptability of different types of evolutionary explanation.

On the most extreme version of the adaptationist view, that held by Alcock, byproduct explanations of female orgasm cannot be evolutionary at all. In order for an explanation to be evolutionary, it must incorporate an account of the adaptedness of the trait. Clearly, this approach amounts to a nonempirical, a priori ruling out of a possible byproduct explanation. Sherman's view has a similar effect, which is that the byproduct view cannot address the adaptedness of the trait. Sherman behaves as if no evidence for the correlation between the trait and the reproductive success of female orgasm has been examined, probably because he finds such a correlation "obvious," even in the face of a lack of evidence. Thus he thinks that questions concerning the current fitness of the trait have not been answered by the byproduct view, and does not understand that the byproduct view does take into account the lack of evidence regarding the relation between the trait and reproductive success.

On either Alcock's or Sherman's argument, the byproduct account is ruled out as an evolutionary account that addresses the supposed evidence of a correlation between reproductive success and the trait of female orgasm. Thus the only explanations that would satisfy the ardent adaptationists are those under which the trait *is an adaptation*. This view begs the question against the by-product view supporters, and it also violates a basic assumption of evolutionary biology, namely, that some traits are not adaptations (Ridley 1996, p. 341).

But in addition to this in-principle ruling out of byproduct explanations, the adaptationists launch a pragmatic argument against the byproduct view. Specifically, the byproduct view is accused of shutting down research into how female orgasm is an adaptation. Expanding their argument, we could argue that it shuts down research into whether female orgasm is an adaptation or not. This seems to amount to the view that we should *never* accept a byproduct explanation of anything, lest that interfere with discovering an adaptive purpose for the trait. Clearly, this cannot be defended in principle, for if a byproduct view is actually correct, we would be best off scientifically to accept it. But there is a legitimate pragmatic worry here about prematurely shutting down inquiry, and I suggest that the worry should be addressed by having the community of researchers pursue two lines of inquiry simultaneously, one into the byproduct view and one searching for evidence that female orgasm might indeed be correlated with reproductive success in some way.

But this pragmatic maneuver aside, we are left with the view that certain evolutionary explanations—byproduct ones—ought *never* to be accepted, first, because they are not adaptive explanations, and second, because they could impede research into adaptive explanations. Thus the byproduct view supporters and the ardent adaptationists seem to have very different views about what could count as a legitimate evolutionary explanation of the trait. I shall return to these issues in Chapter 8. But now I turn to various claims that female orgasm is indeed correlated with reproductive success.

# CHAPTER 7

# Sperm-Competition Accounts

In the previous four chapters, I reviewed both adaptive and non-adaptive explanations for the evolution of human female orgasm. As I argued in Chapters 3 and 4, there are serious evidential problems with the adaptive accounts reviewed there. The available evidence better supports the byproduct hypothesis discussed in Chapter 5, but an open issue remains regarding whether or not there is a connection between reproductive success and female orgasm.

Sperm-competition approaches to understanding female orgasm are the newest on the scene, and they offer the most detailed data yet relating female orgasm to current reproductive success. The sperm-competition accounts available all rely on a claim about female orgasm's role in moving sperm through the reproductive tract. I shall first review this fundamental claim, and then, following a brief overview of the basics of sperm competition, I shall examine several related hypotheses by Baker and Bellis, and Thornhill and colleagues, concerning the evolution of female orgasm.

## The Upsuck Hypothesis

Robert Smith, the first developer of the sperm-competition theories discussed in this chapter, relies on research done by Cyril Fox and his colleagues to establish a connection between female orgasm and

179

the likelihood of fertilization. Claims that the uterus caused suction during orgasm have been widespread throughout the ages. Fox and colleagues (1970) claim to have shown that there is a drop in pressure within the uterus following orgasm. This is believed to promote a sucking mechanism of the uterus at female orgasm that helps move the sperm into the reproductive tract, which is supposed to increase the probability of fertilization upon insemination through intercourse by the male. If the sperm-upsuck theory is correct, then it would constitute the much-sought connection between female orgasm and a feature, fertility, which might be connected to reproductive success, which would thus make it more plausible that female orgasm might serve an adaptive function. (As I shall discuss later, however, an increased fertility rate might not be a reproductive advantage, except in cases of borderline fertility.) In this chapter, however, I conclude that there is in fact no good evidence to support the upsuck hypothesis.

Smith, when explicating his sperm-competition scenario, cites Fox and colleagues as using radiotelemetry to document a drop in pressure in the uterus following orgasm. These results, writes Smith, "confirm the uterus creates suction immediately following orgasm and support the idea that female orgasm may actively transport semen and thereby facilitate fertilization" (1984, p. 643). It is this very idea, as we shall see, that lies at the heart of both major studies examined in this chapter (one by Baker and Bellis, and one by Thornhill, Gangestad, and Comer), so I do well to examine it here.

Consider the claims made by Fox and colleagues (1970). With regard to the drop in pressure, Fox and colleagues wrote, "there followed a sharp post-orgasmic fall in pressure [from +48 and +40 cm $H_2O$] to −26 cm $H_2O$ (Text-fig. 3c) and −16 cm $H_2O$ (Text-fig. 4a) in the two experiments" (1970, p. 247). They obtained these results by placing a remote radio pill, which measured pressure directly, into the uterus, and recorded the results of two episodes of intercourse in a single female subject. Testing whether the sperm were actually moved at all by the hypothesized suction was not part of this experiment. Thus the experiment documented a sharp fall in pressure within the uterus, but no motion of semen was measured or observed. We should also note that the uterus undergoes several regular, baseline contractions per minute in the resting state. This experiment tests uterine pressure, not contractions.

In earlier studies, Grafenberg (1950) and Masters and Johnson (1966) had attempted to document a movement of semen through the cervix as a consequence of orgasm. They found no such movement. Grafenberg's experiment consisted of placing a plastic cervical cap filled with a radiopaque oil over the cervix in an unspecified number of women. The cap was left in place for a whole cycle between menstrual periods, and the women had frequent sexual relations with orgasm. Repeated X-rays at various times never showed that the radiopaque fluid had moved into the cervix or the body of the uterus. All of the radio contrast medium stayed in the cervical cap (1950, p. 147).

Masters and Johnson studied the problem under slightly different circumstances. Like Grafenberg, they placed a radiopaque fluid in a cervical cap. They took six subjects varied in age and in how many children they had had. Because of complaints made about the Grafenberg study, that his radiopaque oil did not resemble semen, Masters and Johnson were careful to use a facsimile of seminal fluid with the appropriate surface tension, specific gravity, and specific density. Masters and Johnson then took X-rays during actual orgasm and also 10 minutes following orgasm. The women achieved orgasm through masturbation. Masters and Johnson reported: "In none of the six individuals was there evidence of the slightest sucking effect on the media in the artificial seminal pool" (1966, p. 123). Nor was there evidence of the radiopaque fluid in the cervical canal or within the uterus. All the women were evaluated at the time of expected ovulation.

Masters and Johnson note that both the measured contractions of the uterus and the mechanics of coitus militate against uterine upsuck. With regard to the contractions of orgasm in the uterus, they note that these contractions start at the back or top of the uterus and go down through the middle, ending at the cervix. In other words, these contractions might have the effect of expulsion rather than ingestion. Masters and Johnson also note that the sperm-suction theory wouldn't seem to work because the upper end of the vagina around the cervix expands during orgasm, and lifts the cervix up away from the pool where the semen would be deposited. Even with these two mechanical problems with the upsuck hypothesis, Masters and Johnson take their own X-ray results as the strongest evidence against the hypothesis.

Fox and colleagues (1970) objected to the experiments done by Grafenberg and by Masters and Johnson, claiming that they were inadequate. Fox and colleagues complain that the Masters and Johnson experiment "suffers from the fact that it was not performed under natural coital conditions, i.e. the stimulation was artificial" (1970, p. 248). This appears to refer to the fact that the clitoral stimulation involved was masturbatory; given the abundant evidence establishing the similarities between masturbatory and coital orgasms, this seems a weak objection at best (Masters and Johnson 1966).

The same objection cannot be raised against Grafenberg, whose subjects underwent orgasm in the course of sex involving coitus. In this case, however, Fox and colleagues object to the use of oils "which may differ substantially in weight and density" from seminal fluid (1970, p. 249). This objection can be raised against Grafenberg, but not Masters and Johnson. In addition, Fox and colleagues objected to the use of a cervical cap, which, they say, prevents the inward and outward flow of matter through the cervix. This raises the question of whether the cervical fluid, or mucus, actually moves outward with the increase in uterine pressure with orgasm, and then inward upon the decrease of pressure afterward. Singer refers to the fact that several physiologists have attempted to observe this directly, but have reported no movement whatever of the cervical fluids during orgasm (Singer 1973, p. 165). Finally, Fox and colleagues emphasize the importance of the pressure on the abdomen by the male; in their experiments, it increased intrauterine pressure by approximately 10 centimeters $H_2O$. ("Centimeters of water" is a standard measure of pressure; 10 centimeters in utero is a significant amount compared to a maximum reading of 46 centimeters and a minimum of minus 26 centimeters.) In addition, the presence of the penis in the vagina alters the state of pressure in the vagina. These would be methodological objections against Masters and Johnson's experiment, but not against Grafenberg's. Despite the presence of these additional pressures from coitus during his trials, Grafenberg still documented no upsuck effect.

Note that the Fox et al. experiment consisted of two trials on one woman. While it is common for physiologists to use a small sample size, this practice is justified only by the assumption that an individual's physiology is typical. Given the very wide range of physiological differences found across women, we must be very wary about whether any conclusions can be drawn about women in general from these data.

Irving Singer (1973) is cited in both the Baker and Bellis and the Thornhill et al. papers as providing support for the upsuck theory. Singer offers a long discussion regarding whether or not uterine upsuck occurs with female orgasm. He acknowledges that Masters and Johnson found that no uterine suction occurred with orgasm, but questions their conclusions on the basis that these women were not experiencing the right "kind" of orgasm for suction to occur. Singer and Singer (1972) hypothesized that there is not one kind of female orgasm, but three: vulval, uterine, and blended. Their basic claim is that convulsive contractions of the vulva and the orgasmic platform are not necessary for female orgasm to occur. Instead, there is another kind of orgasm involving emotional changes and apnea (the momentary, involuntary holding of breath), which is immediately succeeded by "a feeling of relaxation and sexual satiation." "This kind of orgasm occurs in coitus alone, and it largely depends upon the pleasurable effects of uterine displacement" (1972, p. 260). The Singers do note a problem with calling this a "uterine orgasm," since Masters and Johnson had demonstrated that every regular orgasm (the Singers would say "vulval" orgasm) is accompanied by uterine contractions. Though the Singers offer no supporting data, their speculation is that with the "uterine" orgasm, the peritoneum (lining of the abdomen) is stimulated, and the uterus contracts more strongly than in other orgasms.

Singer remarks that the subjects in Masters and Johnson's experiments involving uterine upsuck were under observation, and were not experiencing coitus, and thus may not have experienced the right "kind" of orgasm. His assumption is that the "uterine" and the "blended" orgasm (which includes both vulval and uterine components) are accompanied by uterine suction and that women under observation and during masturbation would be more likely to have purely vulval orgasms. Singer also objects to Masters and Johnson's experiments that showed that the waves of uterine contractions during orgasm "were observed to be expulsive, not sucking or ingesting in character" (1973, p. 162). He notes that the electrode readings that Masters and Johnson took during these experiments are difficult to replicate because the methodology is not described. But he shows no signs of having attempted to design a new experiment showing electrode readings that contradict the expulsive orgasms observed by Masters and Johnson.

Singer also contrasts the results of Masters and Johnson with the work of Fox and colleagues (1970). He claims that "it is only by assuming that human females can undergo more than one type of orgasm that we can explain why it is that Fox, Wolff, and Baker recorded a pattern of uterine contractility quite different from that depicted by Masters and Johnson" (1973, p. 167). Thus we have the introduction of a new type of orgasm in order to support the hypothesis of uterine suction.

Much of the force of Singer's argument comes from comparisons with other species: "If it were the case that uterine contractions during coitus had no effect upon uterine suction and sperm transport, women would differ drastically from most of the lower mammals" (1973, p. 164). He then reviews work on rats, dogs, and cows, in which nonorgasmic uterine contractions have been observed, and have been associated with sperm transport. No studies on primates are cited; indeed, Alan Dixson notes their absence in the literature (1998). Noting that various physiologists have watched the cervix during human female orgasm and found no movement of the mucus in the cervical os (opening)—indicating no suction—Singer once again faults the definition of orgasm being used. "Some orgasms, e.g., vulval orgasms resulting from either masturbation or coitus, may involve no movement of mucus even though uterine and blended orgasms do. These latter orgasms, which the physiologists could not readily have had access to in the laboratory [because they would occur only with a certain style of coitus], might duplicate the patterns of uterine suction found in lower mammals" (1973, p. 166).

Singer does note an interesting correlation between the administration of the hormone oxytocin and uterine contractions (1973, p. 166; Fox et al. 1970, p. 250). Oxytocin is well known as the hormone involved in milk letdown in mammals, including human beings, in which stimulation of the nipple or the presence of a crying baby can induce the breasts to release milk. Oxytocin is also significant in the second stage of labor (Blaicher et al. 1999, p. 125). The connection between oxytocin and uterine contractions occurs in a variety of mammals, but not universally (Overstreet 1983, p. 508; for example, not in rabbits). But a recent set of studies on oxytocin and uterine contractions in human beings may actually provide some limited support for the possible existence of a mechanism that can serve in the place of uterine upsuck.

Wildt and colleagues (1998) found a rapid uptake of radiopaque microparticles suspended in a semen-like substance into the uterus following intravenous or nasal application of oxytocin in women. Wildt's group performed continuous recording of intrauterine pressure before and after oxytocin administration, and found an increase in the strength of uterine contractions after administration, as well as a pressure gradient with oxytocin that might be compatible with suction, in which the pressure is lower away from the cervix than at the cervix.

In their experiments, Wildt and colleagues placed two pressure-recording devices into the uterus, one closest to the fallopian tubes and one next to the cervical opening. The base state, before any hormones were administered, showed that there were steady contractions of the uterus, with a frequency of 3 per minute, and with an amplitude of 10–20 millimeters of mercury (Hg). Immediately following the administration of oxytocin, both the basal pressure and the amplitude of the contractions increased—uterine pressure became greater and the strength of the uterine contractions increased. In addition, there was a difference in pressure between the two recording sites, with a pressure gradient showing between the cervix and the far end of the uterus, wherein the cervical pressure was higher than the pressure near the fallopian tubes (1998, p. 660). These experiments were all done in the absence of orgasm.

While this result regarding pressure gradients might seem to support the results of Fox and colleagues, Wildt and colleagues reject the idea that the data of Fox and colleagues could account for the immediate uptake from the vagina of the marked particles, and their transport thereafter (1998, p. 664).[1] Instead, Wildt and colleagues attribute rapid transport of sperm to the peristaltic (intestine-like) contractions of the uterus—much like the contractions documented in dogs and cows—and to the muscular layers of the fallopian tubes. The relevant peristaltic contractions occur with a frequency of 2 to 5 per minute in healthy women at all times. They also show a characteristic pattern of propagation, depending on the phase of the menstrual cycle. Wildt and colleagues found that a fundocervical (from the back end of the uterus to the cervix) direction predominated during the follicular phase, while during the luteal phase, the peristaltic waves went from the cervix to the back of the uterus (1998, p. 664).[2] Note that these waves occur constantly, and not just when a woman is sexually engaged or is experiencing an increase in oxytocin levels. These contractions can, however, be influenced by a variety of hormones, including prostaglandins, vasopressin, and oxytocin.

Wildt and colleagues note that there are a number of observations in women and in experimental animals that suggest that oxytocin plays an important role in eliciting (nonorgasmic) contractions of the uterus. Moreover, "it has been shown that oxytocin is released from the posterior lobe of the pituitary gland in response to vaginal distension, in response to cervical stimulation and during intercourse in response to tactile as well as emotional stimuli" (1998, p. 664).

All of this is relevant to the orgasm hypotheses because orgasm is known to be correlated with oxytocin release. In addition, there are cyclic changes in oxytocin levels; oxytocin peaks at the time of ovulation and is significantly elevated in the follicular phase as compared to its level during the luteal phase. Oxytocin levels were found to rise significantly in 12 women after orgasm and this might seem to suggest a possible mechanism for the upsuck hypothesis (Blaicher et al. 1999, p. 126). Perhaps the orgasmic increase in oxytocin levels prompts a series of uterine contractions such as those documented by Wildt and colleagues, thus creating a pressure gradient that would conduce to the movement of sperm through the cervix.

The only problem with this uterine-upsuck hypothesis of female orgasm is that, according to these studies, uterine upsuck would be expected to occur frequently and in the absence of orgasm, which makes orgasm unnecessary for any increase in fertilization upsuck might provide. Fox and colleagues admit this very point (1970, p. 250). The fact that oxytocin levels increase significantly with vaginal distention and cervical stimulation implies that the act of intercourse alone will produce a surge of oxytocin that may itself result in uterine contractions of the sort that help sperm transport, with or without orgasm. Sexual stimulation alone is enough to elevate oxytocin levels, also increasing the peristaltic waves or contractions that Wildt and colleagues believe are the causes of rapid sperm transport.

Perhaps the real question lies, then, in whether or not there is some significant difference between the effects of oxytocin with sexual stimulation and with orgasm itself. One of the most striking things about the Blaicher data is the wide range of levels of both baseline and orgasm-related oxytocin levels among the 12 women in the sample. Baseline levels range from 4.9 to 23.6 picograms per milliliter, while the levels at 1 minute after orgasm range from 6.3 to 42.9 picograms per milliliter (1999, p. 126). One woman moved from a baseline of 4.9 picograms per milliter to a post-orgasmic state of 6.3 picograms per milliliter, another went from 8.6 to 8.8 picograms per milliliter, while a third went from 23.6 to 42.9 picograms per milliliter. Thus one woman's orgasmic oxytocin level was approximately one fourth of another woman's baseline level, and one woman moved 19.3 picograms per milliliter, while another moved only 0.2 picograms per milliliter. If there is a dose relation between uterine contractions and oxytocin concentrations, it is hard to know what to conclude from these data. Perhaps the woman with the 6.3 picograms per milliliter orgasmic oxytocin level never experiences a high enough level to enhance the peristaltic contractions of the uterus, or perhaps the woman with the 23.6

picograms per milliliter level experiences robust peristaltic contractions in the absence of orgasm. We just do not know.

Even if we take all the oxytocin and contraction studies at face value, there is no way to tell whether orgasm itself is providing a necessary boost to the peristaltic contractions that are believed to aid sperm transport, or whether sexual excitement and stimulation are already doing the job. If the latter is the case, then the sperm-competition accounts of female orgasm I review in this chapter are superfluous. For if sexual excitement or the stimulation of intercourse are sufficient to provide aid to sperm transport, the female orgasm itself is unnecessary to achieve uterine movement of sperm. It is still conceivable that orgasm could make a marginal difference—that orgasm could provide a crucial added boost to sperm transport if we assume that everything in Wildt et al. 1998 is correct. But the problem is that the evidence that orgasm makes such a difference just is not there. The way the evidence stands now, Wildt and colleauges reject the upsuck theory as advanced by Fox and colleagues, emphasizing instead the role of regular peristaltic contractions of the uterus in fertilization. And this is not enough to establish that female orgasm is an adaptation. It could be that oxytocin release upon orgasm is simply a side effect of the important role of uterine contractions during labor. Furthermore, if a connection between the peristaltic contractions and the female orgasm could be made, and if it could be shown to make a significant difference in fertility, there would still be questions about whether it increased reproductive success. Moreover, even if a contribution to current reproductive success could be shown, it would be insufficient to conclude that the trait evolved as an adaptation (see Chapter 6).

In this context, there are three connections to be made in order to tie together female orgasm and reproductive success. First, orgasm needs to be tied to uterine upsuck; second, uterine upsuck needs to be tied to increased fertility; and third, increased fertility needs to be tied to comparative reproductive success, wherein women who have orgasms are found to have a higher chance of getting their genes into future generations than women who do not. Let me begin with the first point: orgasm needs to be tied to uterine upsuck.

As we have seen, there are conflicting data about this issue. At best, we are at a stalemate. One set of researchers found negative pressure in the uterus after orgasm, suggesting the possible creation of a vacuum or sucking phenomenon, and one set of researchers found positive pressure that suggests instead an expulsive phenomenon. It is unknown whether the increased pressure in the first experiment was enough to move semen through the cervix or uterus. In addition, two sets of researchers found no evidence for movement of semen-like substances during and after orgasm, and no researchers have found evidence of actual upsuck through the cervix related to orgasm. Some movement of particles through the reproductive tract was found in the oxytocin experiments, but the movement was not specifically associated with orgasm—in fact, it was investigated in the absence of orgasm. Thus the evidence is ambiguous at best for the phenomenon of uterine upsuck. After all, three studies suggest no upsuck related to orgasm, and the one study that does consists of a total of two experiments done on the same woman, which document not upsuck itself but a change in uterine pressure. The oxytocin research looks to be the most promising source for evidence of a special upsuck phenomenon related to orgasm, and should be further investigated.

Second, a connection needs to be made between uterine upsuck and fertility. It may seem to be obvious that fertility would increase if more sperm were transported earlier to the site of the egg, but is this true? It needs to be established, and not a single experiment has addressed this issue, to my knowledge. Suppose that the earlier transport of the sperm made no real difference in overall fertility rates? These questions must be answered in order to claim a link between uterine upsuck and fertility. John Bancroft, the director of the Kinsey Institute, notes that there is no evidence "linking orgasmic potential with fertility in the human female" (1989, p. 86). However, I note that fertility clinics routinely encourage women to masturbate to orgasm after artificial insemination, but again, this practice is not based on any evidence I can find. In fact, it is puzzling that the upsuck advocates do not simply study the alleged connection between orgasm and fertility directly. Until such a connection can be established, arguments about the mechanism of the connection are moot.

Third, a connection needs to be made between an increase in fertility and reproductive success. Is more fertility actually better? We are a species that breeds by producing one or two offspring at a time, each of which require intense and long investment in order to reach sexual maturity. The pregnant mother's overall reproductive success—her success in getting her genes into future generations—depends on much more than how many times she can get pregnant. It depends on being able to feed, nurture, and raise her offspring to the point where they can be successful breeders themselves.

Sarah Hrdy has written a comprehensive book on motherhood, *Mother Nature,* in which she details the extent of the investment that must be made by females in order to get their genes into future generations (1999). Women pay for each infant in calories and calcium from both pregnancy and nursing (nursing being usually a 3- to 5-year phenomenon until very recently in human history), in opportunities to invest in other children present and future, and in their own increased risk of death during famine. How could the females afford to raise a child, especially during those intensive first few years? Based on present hunter-gatherer societies, it seems that the only way to do it is by birth spacing: by having a child only once every 3 to 5 years (Howell 1979; Irons 1983, p. 198). Moreover, when infants are born disabled or too soon, the mothers may commit infanticide, resulting in a major loss of energy and resources invested in the pregnancy (Hrdy 1999, p. 184). Thus fertility rates are normally kept very low through breast-feeding the current infant, which triggers a feedback loop that prevents ovulation (Hrdy 1999, pp. 194–195). This pattern of long-term lactation and long birth spacing is found throughout the primates. All this amounts to a serious problem with equating increased fertility with increased reproductive success. In the hunter-gatherer lifestyle of our ancestors, a woman who is more fertile than average is unlikely to benefit her long-term genetic line by that fact alone.[3] Being fertile enough to get pregnant *at the optimal time* is an advantage, but more is not necessarily better. This suggests, by the way, a further possible avenue of research for upsuck advocates—a connection between orgasm rates and pregnancy timing would, if discovered, be a salubrious finding for an adaptive account of female orgasm. No such study has yet been performed, to my knowledge.

Thus we still have a host of problems in trying to link female orgasm to the disputed phenomenon of uterine upsuck and then to reproductive success. The evidence connecting uterine upsuck to orgasm is inconclusive at best, the evidence linking uterine upsuck to fertility is nonexistent, and the evidence concerning birth spacing and reproductive investment seems to cast doubt on the expectation that a general increase in fertility would be linked to higher reproductive success.

Of course, an effect that could help a woman obtain a pregnancy with a genetically superior male when she is at the right point in her birth spacing might be an advantage. This is related to what Thornhill and colleagues (1995) try to argue for. But their view is reliant on Baker and Bellis's evidence regarding uterine suction, which, as we shall see, is very problematic. The evidence that began the lines of inquiry followed in these most recent adaptive accounts for female orgasm rely on the study by Fox and colleagues (1970). Given the challenges facing the reconciliation of Fox and colleagues' indirect evidence with the lack of direct evidence of uterine upsuck, this foundation looks shaky. In addition to the general difficulties with connecting female orgasm to reproductive success, perhaps the most serious problem of all with the experiment conducted by Fox and colleagues is that it involved only one woman, whose orgasms were recorded on just two separate occasions. This is the sum total of their data. There is no reason to believe that this one woman shows a typical response. So it seems especially inappropriate to initiate entire theories of the evolution of female orgasm on the basis of results generated by one woman, as the authors considered in this chapter do. The writers considered here all cite Fox and colleagues' paper as providing definitive evidence for the existence of a basic phenomenon of uterine suction, and build entire theories of orgasm out of that phenomenon.

## Sperm Competition

Geoffrey Parker (1970a, 1970b) was the first to explore the evolutionary implications of sperm competition, which he defined as the competition between sperm from different males to fertilize the eggs of a single female (Birkhead and Hunter 1990, p. 48). Most research on sperm competition has investigated how one male tries to get an advantage for his sperm over that of competing males that also mate with the same female. One mechanism of such sperm competition occurs in dragonflies and damselflies. In these insects, the male's penis is shaped like a scoop or brush. When mating, the male removes the previous male's sperm from the sperm-storage organ of the females. This results in the last male mated being the father of up to 90% of the offspring (Birkhead and Hunter 1990). But the research of interest here involves cases in which the female influences the chance of fertilization of various males' sperm. There have been a variety of mechanisms proposed, but most of them are poorly documented. One of the few patterns well documented in a variety of animals is that "females can influence the amount of sperm received during copulation in a manner that favors particular individuals as sires" (Simmons 2001, p. 317; Singh et al. 2002). For example, Rodriguez (1995) documented that females of *Chelymorpha alternans* (leaf beetles) would sometimes expel a droplet of semen from their reproductive tract after copulation. This ejection of sperm was associated with fewer sperm being stored in the female. (Many female insects have sperm-storage organs, from which they fertilize their eggs upon laying.) Cutting the muscles of the sperm-storage organ changed the probability that the female would eject sperm, which suggests that females can have an influence on sperm storage of particular males, which can then have an effect on the males' likelihood of reproducing (Simmons 2001, p. 289; see Simmons 2001, chap. 9, for an excellent review of female control mechanisms and evidence for them).

Much of the subsequent research on sperm competition has aimed to identify the selective pressures favoring multiple mating by both males and females where it occurs (for insects, see Simmons 2001). There is debate about the frequency of multiple matings by female human beings, but all of the explanations considered in this chapter assume that such matings are common enough to form a significant selection pressure (but see Daly and Wilson 1999). However, there is serious question whether sperm competition occurs among human beings at all. Among primate males, there are certain traits that are associated with species in which the female does copulate with multiple males within a short period, thus setting up the conditions for sperm competition.[4] None of these are found among human males, which suggests that human beings are not, in fact, a species in which multiple-mating females have exerted much selection pressure at all on male traits. Roger Short, one of the founders of research on sperm competition, writes, "We now know that the size of the testis relative to body weight is an infallible indicator of the mating system . . . and there is no doubt that the relatively small 20g testis places us amongst the single:male mating species" (1997,

p. 22). If Short is right (and he is authoritative on the subject), human beings do not undergo sperm competition, and the accounts presented in this chapter are undermined. I shall, however, continue investigating the explanations for female orgasm based on the presumption that sperm competition does, in fact, occur in humans, keeping in mind that it may not.

Of particular interest here is why it might be advantageous for a female insect, bird, or mammal to mate with more than one male, when it takes just one to impregnate her. The major hypotheses are that there are material or genetic benefits, or both, that accrue to a female with more than one male. According to Robert Smith's speculative discussion on human sperm competition, possible benefits might include good genes; fertility back-up; material resources; protection for self and offspring; enhancement of social status; and son's-effect genetic diversity (1984, p. 609). The son's-effect hypothesis is a particular version of the good-genes account of sexual selection. The basic idea is that the male offspring (sons) of sexually attractive males will, by and large, be more sexually attractive than average since they will inherit (to some degree) the traits that made their father attractive. Having sexually attractive sons is a benefit to mothers as well as fathers since both sets of genes will do well as a result of the sons' (expected) success. There is therefore an incentive for females to mate with sexually attractive males outside of their usual partnership in order for their genes to benefit from the future success of "sexy" sons. I focus on the widely used "good-genes" hypothesis, in which mating outside a pair bond is taken to supply superior genetic material. Smith notes that this possible benefit to female multiple mating is "controversial and assailable on theoretical grounds" (1984, p. 610).

Some females who mate (in an otherwise monogamous system) with a genetically low-quality male would benefit by seeking out matings with males with better genes than their present mate. The idea is that the female would obtain resources from her primary mate and then mate opportunistically with males whom she perceived as genetically superior (a system called "facultative polyandry"). Note that this assumes that "human females are capable of discriminating genetically-based differences among potential mates and deciding which phenotypes are superior" (Smith 1984, p. 610; Daly and Wilson 1999). As Smith notes, this assumption is highly controversial.

Orgasm enters the picture with Smith's hypothesis that "it would be adaptive for facultatively polyandrous or promiscuous females to assist the sperm of preferred mates and handicap those of non-preferred mates. In this way a female might exercise some control over the paternity of her offspring while pursuing the nonreproductive benefits of facultative polyandry. Orgasm might permit females to achieve both of these objectives" (1984, pp. 642–643). R. Robin Baker and Mark Bellis subsequently took up this hypothesis—but more on that later.

Smith does acknowledge Symons's views on female orgasm as a competitor, but concludes, "If there is as much variation in the ability of human females to achieve orgasm as is noted by Symons in support of his thesis, variants must have existed upon which natural selection could have operated to remove this complicated 'artifactual, possibly maladaptive trait' from female populations" (1984, p. 643). Here Smith is referring to Symons's entertainment of the possibility that female orgasm might be mildly maladaptive because, according to Masters and Johnson, it might expel semen from the uterus. Smith is quite right to call into question this hypothesis that orgasm might be maladaptive, because surely enough variation in the trait has existed for orgasm to be selected against. However, Smith's argument has no force against Symons's main hypothesis, that orgasm is an embryological byproduct of a male adaptation, which Smith calls "naïve" and "chauvinistic." Under that hypothesis, no selection would have occurred on female orgasm per se, and we would expect the wide variation in the trait that we in fact find (see Chapter 5).

As an alternative to Symons's view, Smith proposes that "human female orgasm may facilitate the transport of semen from the vagina into the uterus, and thereby provide the female with some control over the use of ejaculates" (1984, p. 643). Unsurprisingly, Smith bases his version of the upsuck theory on the work of Fox and colleagues (1970), and claims that these results "support the idea that female orgasm may actively transport sperm and thereby facilitate fertilization" (Smith 1984, p. 643). Smith accepts these findings without noting the existence of contradictory evidence, and goes on to refine the account of sperm manipulation (1984, p. 643). He hypothesizes that the female may have orgasm with a preferred male or by masturbating to orgasm following coitus with a preferred partner, while avoiding orgasm during or following coitus with the less desirable potential father. Smith notes that anecdotal evidence suggests that female orgasm during coitus is more common if the woman "feels generally secure and is convinced of at least some degree of future support from her partner" (Smith 1984, p. 644).

One question that immediately arises is how well this picture fits with the female's seeking of extra-pair male matings. The scenario for good-genes sperm selection would seem to require that the female had more orgasms outside of the pair bond than within it and mostly with males who are genetically superior to the male with whom she is pair bonded. First, are extra–pair-bond copulations usually performed with superior males? Second, if it is security that would encourage her to have an orgasm, it is unclear from Smith's paper exactly why the female should feel more secure in the context of the extra-pair matings. If she is more likely to have orgasm with a partner from whom she expects future support, this would seem to militate against her having orgasm with an extra-pair male. Nevertheless, under the good-genes hypothesis of extra-pair matings, she should have more orgasms outside the pair, when she is likely to become pregnant by the male with the good genes. This is a problem that is unaddressed in Smith's account.

But there is a more serious problem with Smith's hypothesis, and it lies in his reliance on Fox and colleagues' data, despite the ambiguous evidence for the basic phenomenon of uterine upsuck, and the lack of evidence showing that negative uterine pressure, even if it were correlated with such upsuck, occurs in more than one woman. Since it is just this upsuck that is supposed to confer a reproductive advantage by enabling orgasmic women to select sperm, Smith's hypothesis would seem to be only as well off evidentially as Fox and colleagues' problematic results. Nevertheless, Smith suggested that the upsuck generated by female orgasm is an adaptation for sperm manipulation. This hypothesis was taken up and developed much further by Baker and Bellis.

## R. Robin Baker and Mark Bellis's Account

Following Smith, Robin Baker and Mark Bellis's major evolutionary claim is that female orgasms evolved because they were effective at sperm manipulation. In 1993, Baker and Bellis published a pair of papers, one of which (1993b) was concerned exclusively with female orgasm. In it, Baker and Bellis map out the consequences of female orgasm in the full variety of possible circumstances. In other words, they describe the "function"—sometimes evolutionary function, sometimes mechanical function—of female orgasm in many circumstances, including orgasms that occur before, during, and after copulation, masturbatory and nocturnal orgasms, and orgasms of virgins.

Baker and Bellis, apparently unconcerned with the contradictory evidence mentioned above, propose that female orgasm acts as the mechanism for sperm manipulation by creating suction in the uterus, which leads to the sucking in of any material—whether it is semen or vaginal fluids—that is present in the vagina at the time of orgasm. They offer data in support, taken from their own studies done of the fluid that is emitted from the vagina following intercourse, which they call the "flowback." I shall offer detailed comments on their data collection and interpretation later; for now, I wish only to chart their argument.

Baker and Bellis's basic claim in the paper (1993b) is that "copulatory and inter-copulatory orgasms endow females with considerable flexibility in their manipulation of inseminates." They suggest a parallel between female insects, who are known to scoop out sperm from their sperm-storage organs, and women, who experience a flowback of seminal fluid following copulation. Their experiments constitute the first attempt to determine the proportion of sperm ejected and retained by human females following normal copulation (1993b, p. 887). They investigate the variation in the proportion of sperm ejected or retained in relation to the socio-sexual situation and in relation to female orgasm. The hypothesis being tested is that "the timing of orgasm, both during and between copulations, is the key feature of the mammalian female's armoury in male:female conflict and cooperation within the female tract" (1993b, p. 888). More specifically: "We propose that nocturnal, masturbatory and copulatory orgasms are the primary mechanisms by which the female influences the ability of sperm in the next and/ or current ejaculate to remain in, and travel through, her reproductive tract. We predict that by altering the occurrence, sequence and timing of the different types of orgasm, the female can influence both the probability of conceptions in monandrous [one-male] situations and the outcomes of sperm competition in polyandrous [multimale] situations" (1993b, p. 888).

Baker and Bellis's study has some real advantages over previous studies. Specifically, it acknowledges the various situations in which female orgasm occurs, and acknowledges wide variations in the timing and frequency of female orgasm in relation to copulation. Thus it is superior to many previous hypotheses because it attempts to account for the known phenomena of female orgasm as it actually exists, rather than as researchers imagine it to be.

One of the four sources of data for their paper is a mail-in magazine survey from a United Kingdom nationwide magazine, *Company*. Out of an estimated circulation of 439,000 copies, Baker and Bellis received 3,679 replies, a 0.84% response rate for the entire readership. Since we do not know how many of the readers were women, it is impossible to estimate the actual female response rate. The respondents answered 57 questions on their sexual behavior, including information regarding their last copulation and their last noncopulatory orgasm. The data included information on 2,745 in-pair copulations and 126 extra-pair copulations.

The confusing nature of Baker and Bellis's survey results makes it difficult to compare them with the scientific sexology literature, for example, with respect to the estimated orgasm rate during copulation. In the Baker and Bellis survey, of experienced women (those who had over 500 lifetime copulations), 84% had experienced orgasm during copulation (1993b, p. 894). This number is affected by Baker and Bellis's odd definition of "copulatory orgasm," which includes orgasms from foreplay and postplay. Their definition says that copulatory orgasm occurs "through self, manual, or penile stimulation as part of a copulation episode (orgasm occurring during foreplay, postplay, or copulation itself)" (1993b, p. 888). Because of this peculiarly wide definition, it is difficult to compare their results with the usual sexology literature. In any case, as one reviewer noted, there is an acute difficulty in considering their survey results to be representative of anything, given their methodology, the apparently low response rate, and the likelihood of a skewed sample (Short 1997).

The other sources of data for their study include (1) sperm counts in whole ejaculates collected by condom during copulation, (2) sperm counts in flowbacks from females following copulation, and (3) the female subjects' subjective estimates of flowback volume. From these data, estimates were made of the number of sperm retained by the female, which were in turn based on estimates of the number of sperm inseminated into the female. Whether and when female orgasm occurred were also recorded along with the flowback data for each act of intercourse.

From their flowback data, Baker and Bellis conclude that whether a female has a copulatory orgasm or not (under their wide definition) has no significant influence on the residual number of sperm retained. However, for female orgasm occurring anywhere from 1 minute before ejaculation to 45 minutes after ejaculation, they find a correlation with high sperm retention. This is compatible with the idea that female orgasm produces a sucking force in the uterus; when there are sperm present, they can be sucked into the uterus and cervix. If sperm were not present, as would be the case in all orgasms occurring more than 1 minute before ejaculation, then there would not be sperm sucked up and held within the uterus and cervix. Thus they conclude that the timing of the female orgasm makes a large difference in the number of sperm retained by the female.

In addition to exploring the timing of orgasm, Baker and Bellis investigated various combinations of events that occurred between one copulation and another, and concluded that having a non-copulatory orgasm between copulations affected sperm retention for the second copulation. They theorize that there are four regimes of sperm retention, organized according to time between copulations and whether and when there was an inter-copulatory orgasm. These four regimes produce three different levels of sperm retention, ranging from very low sperm retention (level I), to high sperm retention (level III). Thus the female is taken to have the means available to her to counteract efforts toward sperm competition by the males.

Baker and Bellis then use their framework with its three levels of sperm retention to analyze their data on extra-pair copulations from their questionnaire survey. They inferred from the survey data that sperm-retention levels were significantly higher during extra-pair copulation than in in-pair copulation. This is just what would be expected if females were mating outside the pair in order to obtain good genes. Baker and Bellis note, however, that they cannot make any strong conclusions regarding extra-pair copulations; their initial conclusions are calculated on the basis of the (probably nonrepresentative) survey data, and they do not have any flowback data from women mating with more than one man.

Baker and Bellis do want to draw several conclusions from their flowback data, however. First, they claim that their flowback data "provide the first direct evidence in support of the 'upsuck' hypothesis proposed by Fox et al. (1970)." Second, they hypothesize that the upsuck mechanism applies to inter-copulatory orgasm as well. However, "The function" of inter-copulatory orgasm, according to Baker and Bellis, "is to slow down or prevent further passage of sperm already present in the cervix, promote a build up of cells and debris in the cervical channel, and thus to reduce sperm retention at the next copulation" (1993b, p. 905).[5]

Finally, in the absence of copulation (such as in virgins) and during pregnancy "the function of nocturnal, masturbatory, and other non-copulatory orgasms could be as an 'antibiotic' mechanism aimed at combating cervical infection" (1993b, p. 905). The relevant mechanism would be the upsuck of the acidic vaginal fluids, creating a hostile environment in the cervix for bacterial and fungal infections.

Thus Baker and Bellis want to go beyond seeing female orgasm as a simple assistant to transporting sperm into the uterus. They consider noncopulatory orgasms, pre-ejaculatory orgasms, and copulations without orgasm, which were ignored in most previous discussions of the evolution of female orgasm. Each of these conditions has a function, according to Baker and Bellis. According to their data, some orgasms reduce sperm uptake at the next copulation, while other orgasms increase sperm retention in other circumstances. They conclude, "The occurrence, pattern and timing of female orgasms emerge from our analysis as part of a female strategy to influence sperm retention from any given copulation" (1993b, p. 906).

What Baker and Bellis have done is describe each outcome of a female orgasm—whether it is retention or nonretention of sperm, or uptake of acidic vaginal fluids—as having a particular function, and each function is thought to be beneficial to the female. Thus when a woman experiences orgasm during pregnancy, the benefit to her is that it serves as an antibiotic. Or when she has an orgasm well before ejaculation, it is a benefit for her to *not* retain the sperm of that male. Surprisingly, Baker and Bellis even conclude from their investigation of flowback during monogamous copulatory and masturbatory orgasms that "both male and female partner[s] may benefit from improved chances of viable conception if *fewer* sperm are taken into the female tract" (1993b, p. 906, my emphasis; 1993a). Thus counterintuitively, if the woman is mating with only one male, the "favored strategy" to conceive is for the female to reduce the number of sperm retained. In any case, they hypothesize one benefit or another for each possible circumstance of orgasm, with no orgasmic or copulatory energy being spent for evolutionary naught. Women are thus seen as using a variety of "strategies" of orgasm in order to manipulate sperm within their tracts. Some women use a "mixed strategy," involving copulatory and inter-copulatory as well as lack of orgasm; some women use a "no-orgasm" strategy; while the others use a strategy involving orgasms, but not during copulation—Baker and Bellis call this the "inter-copulatory orgasm dominated" strategy.

But there seems to be a fatal flaw in Baker and Bellis's account. According to them, no matter which strategy a woman uses, "*all* are capable of manipulating sperm retention across the whole range from level I to level III" (1993b, p. 908; my emphasis), *including* the "no-orgasm" women. By their own argument, then, sperm manipulation cannot provide a selection pressure on the female to produce orgasm of any kind, since, according to them, the same levels of sperm manipulation are available to totally nonorgasmic women. Indeed, Baker and Bellis's results regarding the various outcomes of females having or not having orgasms at particular times and in particular ways thus cannot be used to support an evolutionary account of female orgasm at all, according to their own conclusions. It is as if an argument had been given of the following form: function X is important to the reproductive success of an organism; trait A performs function X, but the organism also performs function X in the absence of trait A. Therefore, trait A exists because it evolved by natural selection to perform function X. Put this way, the logic of the Baker and Bellis paper is transparently faulty. The fact that Baker and Bellis still maintain that sperm manipulation provides an evolutionary pressure on female orgasm (1995, p. 50) therefore remains thoroughly puzzling.

A further problem arises from a later analysis they give of their own data. According to their 1995 book, the upsuck of sperm following orgasm actually does not appear to work in the period from 1 minute through 10 minutes after male ejaculation (1995, p. 236). They found significantly lower levels of sperm retention during this period, comparable to those of the "low-retention" orgasms that precede ejaculation. They speculate that the reason for this is the coagulation of the seminal pool. They nevertheless divide female orgasm associated with copulation into just two regimes of sperm retention, even though their data show low-retention orgasms before ejaculation, high-retention orgasms within 1 minute of ejaculation, *then* low-retention orgasms for the following 10 minutes, then high-retention orgasms again after that. So, whereas Baker and Bellis claim that pre-ejaculatory orgasms are low-retention, and post-ejaculation orgasms are high-retention, their own data in fact show a much more varied pattern. If, indeed, the hypothesized upsuck mechanism of female orgasm does not work during the co-agulation of the seminal pool, as they suggest, this leaves a fair number of potential female orgasms—those occurring anywhere from 1 to 10 minutes after ejaculation—in the low-retention regime (about 10% of their own data points fall in this period). Thus their generalizations regarding post-ejaculatory female orgasms are ren-dered doubtful by their own data.

Perhaps Baker and Bellis have overestimated the facility with which females can manipulate sperm. Can we not still extract evi-dence for the upsuck hypothesis from their data? The answer is "no," and the reason lies in the data themselves. There are such se-rious problems with the fundamental data set on flowbacks used by Baker and Bellis that it fails to meet basic scientific standards of evi-dence. I now turn to these deeper problems with their data.

In one data set, they have 1 out of 11 couples in the sample con-tributing 93 out of the 127 data points (73% of the data). Four of the other 10 couples contributed 1 data point each, while another 3 couples contributed a total of 7 points, or 5.6%. Extrapolating to the population at large based primarily on the results of a single subject violates standard statistical practice. In another case, 2 of 5 males in the sample contributed 33 out of 40 data points. The chief problems with samples such as these are that they are too small, and that they cannot be subjected to statistical tests that assume a normal distribution (the classic bell-shaped curve) because of their skew. Baker and Bellis do recognize this second problem; they write, "to avoid making any assumptions concerning the normal-ity of our data, we calculated probability values using only non-parametric statistical tests" (1993b, p. 889; see the box on page 206). They claim that they do not use, except in two instances, sta-tistical tests that require a normal distribution, but this is not true. Throughout their paper, they use such normal-based statistical

## Statistical Problems in Baker and Bellis

Baker and Bellis claim that they "calculated probability values using only non-parametric statistical tests" (1993b, p. 865). In fact, Baker and Bellis use the statistic L defined by Meddis (1984, pp. 25–31) but then standardize L to get $z = L - E(L)/\partial L$ and use the normal approximation to test whether z is significant. In most cases their samples are extremely small (for example, in their paper on female orgasm, we see $N = 11$ for collected flowbacks or $N = 9$ for estimated volume of flowback), and so the departure from normality must be regarded as a problem. Despite their claim that "our conclusions should not be questioned on grounds of skewed sample sizes and undue bias from the more prolific females," this bias cannot be ignored. One couple (B in table III of 1993b) provides 93 out of 127 flowback reports. Yet this couple's median receives the same weighting as that of the four couples who contribute only one report. On page 893 (1993b) data are reported for a single female together with a significance value from the normal distribution. In other places medians of one or two observations are used with the same authority as medians of larger samples (for example, 1993b, table III; 1993a, table V).

The tiny sample problem is important when testing the difference of two correlations. Apparently (1993a, p. 871), Fisher's z transformation was used on Spearman's Rank correlations even though the sample sizes were only 20 and 8, considerably below the usually accepted sample size for Fisher's z.

In general, the vagueness of the definition of L as used to compute z constitutes a problem. Sometimes the values of $\{l_i\}$ that maximize L are used without any statement as to why these $\{l_i\}$ provide the correct test statistic. The values of $\{l_i\}$ chosen always appear to confirm the authors' preconceived notions about the ordering of their conditions.

methods, suggested by Meddis (1984), to approximate their non-parametric statistics, and to demonstrate the statistical significance of their results a total of 60 times. However, for the normal approximation to be valid, a large sample would be needed.

Just as problematically, Baker and Bellis seem sometimes to handpick their samples to show specific results. For example, they start off with a full sample of 32 couples that participated in the studies on flowback. Yet when particular tests are performed on the sample, they sometimes choose to use 11 of the couples with 127 data points, sometimes 18 with 230 flowbacks, or sometimes 7 couples with 117 flowbacks, usually without explanation. In fact, a total of 37 different combinations of couples and data points are used throughout the paper on female orgasm alone. Some of this makes sense, because the tests they want to do often demand dividing the sample to exhibit the relevant characteristics, for example, when they want to test women who have menstruated between two copulations against those who have not (1993b, p. 900). But there are also cases in which no such rationale is obvious or warranted. For example, they do not explain why, out of the 34 male-female couples with records on 323 in-pair copulations, they use only 84 of the flowbacks donated by 24 couples. It might appear that they are just removing the estimated flowbacks and only using the measured ones, but elsewhere they use 127 flowbacks contributed by 11 females. And the selective use of data continues. Shortly after establishing the 127 copulations by the 11 females as their basic data group, they select 8 females with a combined 103 flowbacks to test a hypothesis about the volume and timing of flowback, and then immediately use 18 females and 230 flowbacks to test a different claim about the number of sperm inseminated compared to the time to flowback. Later, they use 7 females and 117 flowbacks, rather than the original data on an earlier page. The worry is that they give no rationale for choosing to use one subset of the samples or another for each particular test. And take table III (1993b, p. 892), in which 1 couple out of 11 contributed 93 out of the 127 data points. When only 8 females, representing 103 flowbacks, are used in a later test, which couples' data points have been thrown out and why (1993b, p. 893)? They obviously left in the female who contributed 93 data points, but who did they leave out?

In sum, the problems with their skewed samples and small sample size, their handpicking of subsamples, and their use of statistical tests requiring a normal distribution call every one of their conclusions into serious doubt. They claim, in the discussion section of the paper, that they have protected the validity of their findings by keeping to nonparametric (nonnormal) statistics, but this is patently not the case when such statistics have been used throughout the paper 60 times. The small sample size is also fatal; nonparametric statistics require larger sample sizes than parametric ones. Thus it is not an exaggeration to conclude that their interpretations of their data remain virtually worthless. They simply cannot be considered as supporting Baker and Bellis's (or any other) larger theoretical conclusions.

Baker and Bellis's conclusions that they have provided evidence for uterine upsuck, and have shown that female orgasm is an adaptation, are thus undermined. They have not, in fact, improved the case for uterine upsuck, and the additional ties needed to conclude that female orgasm is correlated with reproductive success are not provided. No evidence is offered that connects uterine upsuck with fertility, and neither is there evidence that fertility is tied to reproductive success.

Now recall that John Alcock, in an attack on Gould's treatment of female orgasm, writes, "the fact that female orgasm apparently can draw sperm closer to the cervix, and thus potentially affect the fertilization chances of an egg, is evidence of the kind of complex functional design that demands adaptationist analysis" (2000, p. 3). This "fact" is supposed to be demonstrated by Baker and Bellis's work (Alcock 2000, p. 5). As should be clear from the criticisms listed above of Baker and Bellis, Alcock's appeal to this research to document a connection between female orgasm and fertility is severely misguided. Alcock also appeals to Thornhill and his coworkers' development of alternative hypotheses about "the possible current utility of female orgasm." Let me then examine this work, which also centers on sperm competition, to see if it provides any more persuasive support for Alcock's position.

### Randy Thornhill, Steven Gangestad, and Randall Comer's Account

Randy Thornhill, Steven Gangestad, and Randall Comer (Thornhill et al. 1995) also present a study based on the ideas of sperm competition laid out by Smith (1984). They pursue Smith's idea that females may be manipulating sperm so as to favor genetically superior males. They assume that mating outside of the pair is likely to involve males genetically superior to the female's usual mate, predicting that these extra-pair matings will involve more orgasms than in-pair matings. They use the upsuck theory of female orgasm to conclude that relatively more fertilizations will result from the extra-pair matings. They use bilateral symmetry measurements of the males as an indicator of male genetic quality. The more physiologically symmetrical the male is, the higher his genetic quality is supposed to be, because symmetry could indicate his resistance to pathogens such as parasites and environmental toxins. Thus Thornhill and colleagues conjecture, if (1) high symmetry is a mark of male quality and (2) female orgasm is a conditional response to male quality, then female orgasm should be associated with male symmetry. It is this prediction that they aim to test in their study.

Thornhill and colleagues (1995) performed a study of 86 heterosexual adult couples; they measured the symmetry of the males and administered a questionnaire to determine the rates of female orgasm and other possibly confounding variables. Male symmetry was calculated by measuring the left and right sides, and included measurements of the feet, ankles, hands, wrists, elbows, and ear length and width.[6] Female orgasm rates were calculated by asking both the male and female partners to give the percentage, in 10% increments, that the female experienced orgasm. This rate was broken down into periods of before, during, and after sexual intercourse, and then additionally divided into periods within intercourse of before, during, and after male ejaculation. They also asked what percentage of the time the female faked orgasm. They calculated the final orgasm figures by averaging the male and female answers, after removing the cases with the most serious gaps between the male and female responses.

Thornhill and his coauthors found that the average rate of female orgasm with copulation was 60 +/− 29.5% of the time. Of these copulatory orgasms, 28 +/− 20.8% occurred before male ejaculation, 14 +/− 11% during ejaculation, and 19 +/− 17.4% after ejaculation. (Note that these data are quite different from the data from Baker and Bellis, who found a majority of female orgasms occurring before ejaculation; 1993b, p. 896.) In order to test their hypothesis that women would have more orgasms with more symmetrical males, Thornhill and colleagues performed a series of multiple regression analyses. They found that only male symmetry significantly predicted female orgasm during copulation, as they had hypothesized.[7] They also tested whether the orgasms during or after ejaculation—those, according to Baker and Bellis, that are supposed to be high sperm-retention orgasms—varied with male symmetry. They found that high men's symmetry significantly predicted the hypothesized high sperm-retention orgasms, but low men's symmetry did not significantly predict the hypothesized low sperm-retention orgasms. Because of this lack of correlation between low-symmetry men and low-retention female orgasms, they fail to show a significant difference between the overall effects of men's symmetry on the retention rate of female orgasms (1995, p. 1610). They note, however, that their study may be insensitive, because they did not separate out the female orgasms occurring 1 minute or less before ejaculation.

Despite the lack of significant difference mentioned above, Thornhill and colleagues conclude that "our findings support Smith's (1984) general notion that female orgasm evolved as a means by which women manipulated sperm competition occurring as a result of facultative polyandry" (1995, p. 1610). This is a real stretch, considering that Thornhill and colleagues' tests involved no extra-pair matings whatsoever, and thus no sperm competition; all of their measurements were done within the pair. Recall that the connection is supposed to be that individual females would be hard-wired somehow to have more orgasms with men who are more highly symmetrical—and therefore genetically higher quality—than with the men with whom they are pair bonded. This experiment provides data regarding female orgasm rates associated with matings only *within* a pair. The result would then have to be extended to extra-pair matings, or else it would not involve sperm competition at all. Hence, it is highly misleading to say that this experiment tests the hypothesis "that orgasm is an adaptation for manipulating the outcome of sperm competition resulting from facultative polyandry" (1995, p. 1601).

The compatibility of Thornhill and colleagues' findings with Baker and Bellis's is also a bit worrisome. Thornhill and colleagues claim that "the pattern of our results is consistent with Baker and Bellis' (1993[b]) specific hypothesis that timing of the orgasm plays a role in the manipulation, although we cannot offer these data as strong support for that claim" (1995, p. 1610). If, in fact, as Thornhill and colleagues claim, higher symmetry is associated with more orgasms overall, this would seem to make the fine-tuning of timing on which Baker and Bellis spend so much effort beside the point. The fact that Thornhill and colleagues do not get a significant difference between low and high sperm-retention orgasms (defined in terms of their timing) would seem to undermine the efficacy of the uterine suction mechanism that is supposed to be doing the work for Thornhill and colleagues. The additional problems, mentioned before, about the tenability of the upsuck phenomenon apply here, as well. Finally, the most important assumption of the entire project is left unsupported, namely, that female orgasm is associated with increased fertility or reproductive success.

There is a further concern with Thornhill and his colleagues' hypothesis. Specifically, the argument is supposed to be that the capacity for orgasm in the female is selected because it increased her relative fitness by allowing her to favor higher-quality males as sires. This mechanism seems to rely on the existence of variability in the female's response to intercourse depending on the quality of the male. But what of the majority of females, who either always have orgasm with intercourse, or who never or rarely have orgasm with intercourse? It seems that the hypothesis by which female orgasm is an adaptation does not apply to them. Nevertheless, the overwhelming majority of these women are orgasmic. Thus we seem to have here a hypothesis about the selective benefits of orgasm as it shows up in a minority of women. If orgasm were really selected as an indicator of comparative male quality, why wouldn't all women be such that they sometimes have orgasm with intercourse and sometimes do not? No countervailing selection pressure is discussed here. Thus the account seems to make little sense.

But there are yet further difficulties with this paper; there are two primary problems with the statistical methods used. The first involves the implausible justification given for including male reports of female orgasm.

Each of the individuals in the couples was asked questions about actual female orgasm during intercourse. Each partner was asked the percentage of the time, in 10% increments, that the female partner experienced orgasm during any sexual bout with the partner. Questions were included about the timing of orgasm as well, concerning whether the female had an orgasm during sexual intercourse in particular, and if so, whether it was before, at the same time as, or after the male's orgasm. In addition, each partner was asked how often the female partner faked orgasm.

The obvious question here is: Why ask both the males and the females whether and when the female experienced orgasm? Aren't the females the best reporters of this information? Thornhill and colleagues claim, "Using both partners' reports of orgasm increases the validity of our estimate" (1995, p. 1605). "To the extent that each partner's report is somewhat valid (as reflected by the sizable correlation between partners' reports) yet fallible, the composite measure should be more valid than either measure alone (Anastasi 1988)" (1995, p. 1605). But they found only a moderate correlation between women's and men's reports on the various questions: Pearson's r ranged from .52 to .24 for the different questions.

In the discussion section of the paper, this curious practice of combining the male and female answers to questions about female orgasm is justified further. Thornhill and colleagues begin by acknowledging that self-reports may contain errors, but claim "our use of reports by both partners allowed us to examine the validity of individual reports. Women's own reports and the reports of their partners correlated substantially (about 0.6)" (1995, p. 1610). Note that this correlation is actually relatively low. Nevertheless, the authors continue by claiming that each set of reports (the men's and the women's) must be "substantially valid" because of this correlation. "Moreover," they continue, "if each set of reports is nearly equally fallible, the average of the two reports should be more valid than either report alone" (1995, p. 1611).

How can men's reports of their partner's orgasms versus women's reports of their own orgasms be equally fallible? Even Thornhill and colleagues' own data undermine this assumption. They discovered that there were numerous cases in which women reported substantial faking that their partners did not detect. Their solution to the problem that men were not good at telling when women were faking was to remove the men's data points when "the proportion of total copulations the woman claimed to fake orgasm was more than 20% higher than that reported by the man" (1995, p. 1605). In other words, if she reported faking 50% of the time, and he thought she faked only 30% of the time, his data were left in the analysis. In all, 10 of the 86 males' reports had to be thrown out; that is a full 12% of the males in the study. And the remaining males could differ by as much as a fifth from the women's reports, and their data would be left in the analysis. (By the way, it is only by cleaning up the data in this way that the correlation coefficients could be moved up to the approximately 0.6 that they later use in their claims.) Finally, they found that males had a systematic bias in overestimating the percentage of female orgasms, which casts further doubt on the reliability or relevance of male estimates of female orgasm (1995, p. 1606; Hamilton 1929).

Thus Thornhill and colleagues' approach, in which the males' and females' reports are treated as "nearly equally fallible" makes little sense, and is certainly not supported by their own data. This brings us to a larger question; why did they want to include those male reports in the first place? As they admit, retrospective studies are always suspect. If they wanted a better estimate of orgasm rate, they should be using daily self-reports from the women, not retrospective data at all. Moreover, it is widely known among sex researchers that it is frequently difficult for a male to discern when a female has an orgasm, so including the male guesses at when the female had orgasm makes no sense. Perhaps the male data are there to beef up the statistics in some way; they do claim to have run the regressions using the female and male orgasm data separately, and to have found a significant effect of male fluctuating asymmetry, of how symmetrical the male's limbs and ears are (1995, p. 1606), but we cannot verify this claim because we are not given the raw data. In any case, treating the two sets of reports as on a par in their analysis brings the entire set of regression studies into question. The male symmetry measures and other variables are being compared to a combined number (taken as a mean) of the male and female reports on female orgasm; hence the basic results of the study depend on the legitimacy of the combined number. Finally, I would note that the self-report reliability tests that Thornhill and colleagues do on their numbers are inappropriate; they are actually comparing apples and oranges: data of self-reports from the females and data regarding an externally observed phenomenon from the males, not self-reports.

There is also another problem with the statistics. Thornhill and colleagues report that the 86 women in their study claimed to have an orgasm during copulation 60+/− 29.5% of the time. They then divide this figure up on the basis of the timing of the orgasm. First, note the very high standard deviation. But the real problem is that they do not publish a distribution of the percentages of women having orgasm with copulation, yet they compute the regression on just this distribution of percentages. In other words, the regression analyses that form the basic conclusions of their paper can legitimately be run on data that are normal only, data that fit the classic bell-curved distribution; but we are given no information about the shape of the distribution here, and percentages are very often *not* normally distributed. Because we cannot see whether the data on orgasm are normal, we cannot tell whether the tests Thornhill and colleagues do on percentages are legitimate. The significance levels resulting from their regression analyses are not very high; thus it could be that by making the data appear more normal, they will get greater significance. Moreover, if we take the raw data from Baker and Bellis (1993b), we can see that their orgasm percentage information was, in fact, not normal. Thornhill and colleagues have not even tried to convince anyone that their data are normal, and thus that the regression analyses they run on these data are legitimate.

Robert Montgomerie and Heather Bullock's recent attempt to

replicate the Thornhill findings makes the situation for Thornhill and his colleagues worse. Montgomerie and Bullock repeated the protocol of Thornhill and colleagues with more than 80 university student couples, using a double-blind measurement procedure. They found "no hint of a relation between any aspect of partner [asymmetry] and orgasm rate in females, despite the fact that orgasm rate in our sample varied from 0–100% and interindividual variation in [asymmetry] was considerable" (1999, p. 62). Further details of the experiment are unavailable, and we do not know whether Montgomerie and Bullock's statistics are subject to the same objections as are Thornhill and colleagues'. Thus in addition to the statistical and theoretical problems, reviewed above, there also seems to be a problem replicating the results of Thornhill and colleagues.

## Conclusion

In sum, the data for the two sperm-competition hypotheses regarding the evolution of female orgasm suffer from serious statistical and theoretical shortcomings. In addition, clear evidence for the necessary mechanism of uterine upsuck is still lacking, and there is positive evidence demonstrating a lack of uterine upsuck. In addition, each of the theories considered here involves an ornate set of assumptions about mating preferences, none of which is adequately supported. I conclude that even a weakly supported adaptive hypothesis for the evolution of female orgasm has yet to be provided.

Yet both the Baker and Bellis and the Thornhill and colleagues studies are taken to be authoritative by many readers. Jolly (2001, pp. 90–91), Rodgers (2001, pp. 323–331), Fausto-Sterling and colleagues (1997, pp. 404, 416), Singh and colleagues (1998), Buss (1994, pp. 75–76), and Barrett and colleagues (2002, pp. 114–115) all appeal to the Baker and Bellis conclusions as if they are fact. The serious statistical shortcomings of the studies seem to have gone virtually unnoticed by nearly all commentators (but see Short 1997; Birkhead 2000, pp. 23–29). How can this have happened?

I suggest that these authors who cite the Baker and Bellis and Thornhill and colleagues conclusions found the uterine suction hypothesis especially well suited to their adaptationist commitments. In order to give an adaptive account of female orgasm, there must be some connection between female orgasm and reproductive success. The hypothesis of uterine suction provides part of such a connection, by possibly linking female orgasm with an increased likelihood of fertilization. I say this is only part of such a connection because much more is involved in reproductive success than an overall increase in fertility, as reviewed earlier. For instance, the female must also have sufficient nutrition to carry out the pregnancy, and must be in a position to raise the infant to a fertile adulthood. Still, any link between female orgasm and fertility—if one can actually be demonstrated—might provide a crucial component of an adaptive account of female orgasm.

Thus there remains the very serious problem, facing all those who want to give adaptationist accounts of female orgasm, that there is only the one, contested experiment connecting female orgasm with any component of reproductive success. Baker and Bellis and Thornhill and colleagues chose to take the data of Fox and colleagues at face value as demonstrating uterine upsuck, despite the fact that Fox and colleagues did not actually experiment on uterine upsuck.

The Baker and Bellis paper, however, claimed to show confirming evidence for the upsuck hypothesis that was independent of the results of Fox and colleagues. Indeed, Baker and Bellis offer an avalanche of statistics supposedly showing the existence of uterine suction. However, as we saw, there are intractable problems with the Baker and Bellis data, ones that make their statistical interpretations worthless. Unfortunately, Thornhill and colleagues also rely on the Baker and Bellis data to demonstrate the existence of a mechanism for upsuck that would enable women to favor sperm from symmetrical men. Thus the two papers collapse like a house of cards as a result of the failings of the Baker and Bellis data.

There are other problems as well. As mentioned before, Baker and Bellis claim that one can get the same results of different levels of sperm retention without orgasm at all. But this means that their account provides no selection pressure for female orgasm, and thus cannot be considered an adaptive account at all, despite their presentation of it.

In addition, Thornhill and his colleagues' experiments do not address the strong conclusions they wish to draw, "that orgasm is an adaptation for manipulating the outcome of sperm competition resulting from facultative polyandry" (1995, p. 1601). There was no sperm competition in their experiments, nor was there facultative polyandry. They simply did not test this hypothesis.

Notice also that there are a string of assumptions made in these sperm-competition accounts, all of which need to be considered when one weighs the evidence. For example, a certain frequency of women copulating with more than one man within a short time period is assumed, although there is little reliable evidence to show that such timing is common enough to provide selection pressure (but see Daly and Wilson 1999). Moreover, human sperm competition was supposed to play a role in the adaptive significance of female orgasm. But whether or not sperm competition itself occurs with any frequency among human beings is itself under doubt, as mentioned before. The key male traits associated with sperm competition in primates do not occur among human males. Thus these theories may not even get their most basic assumption off the ground. Without male sperm competition, the suggestion that female traits, such as female orgasm, are supposed to result from conditions of multiple matings is seriously undermined.

In addition, there are the assumptions that females seek out matings with males with superior genes, that they can detect such males, and that individual women are more likely to have copulatory orgasms with more symmetrical males. The fluctuating asymmetry of the males is a stand-in, in the hypothesis of Thornhill and colleagues, for male genetic quality. Thornhill and others have pursued this notion, that women prefer more symmetric males, in a series of experiments. The conclusion of this paper, however, that women are more orgasmic with symmetric males, currently has the alleged support of their own (statistically problematic) findings, but is challenged by a replication study undermining it. Further research is needed before any conclusions can be drawn.

Thus the status of the sperm-competition hypotheses considered in this chapter is considerably less firm than many of its proponents appear to believe. The hypothesis of uterine upsuck is virtually unsupported, and it is the linchpin of the mechanisms suggested. Here, it seems, yearning for an adaptationist account has led researchers to accept the existence of a special connection between female orgasm and reproductive success, where that connection is as yet unfounded. The hypotheses in this chapter are good examples of what can happen when a researcher's primary commitment is to finding an adaptationist account of female orgasm. I shall examine in the next chapter how easily the blind reliance on such deficient studies can occur.

# CHAPTER 8

# Bias

I have reviewed and critiqued all of the evolutionary explanations for female orgasm that I could find. The goal of this chapter is to draw some more general lessons from the material I've presented in the previous chapters. Are there common causes for the evidential failures I have documented? Is there an explanation for why Symons's hypothesis, with its supporting evidence, has been rejected in favor of hypotheses with fatal statistical problems? To address these questions, I undertake three levels of diagnosis. First, I review some of the frequently appearing evidential problems with evolutionary accounts of female orgasm. Second, I suggest four background assumptions that are playing crucial roles in the production of these evidential problems. Finally, I examine the commitments of the community of evolutionary theorists involved in the debates about female orgasm. I argue that using Helen Longino's account of how scientific communities produce objective scientific results can help us understand what went wrong in this community (and in its analysis of Symons and of Baker and Bellis). In addition, this case serves as a source of good fit for both Longino's and Elizabeth Anderson's models of scientific bias. In contrast, Philip Kitcher's more traditional account (1993) and Miriam Solomon's more psychologistic account (1995) fail to locate what is wrong with these adaptationist accounts. I conclude that the evolutionary community has failed to critically engage a key issue about adaptation because of the varying commitments of different theorists, but I suggest how the situation might be improved.

## Problematic Assumptions

Using minimal standards for evolutionary explanations, I have emphasized that a number of the assumptions made by various practitioners in explaining female orgasm have been undermined or unsupported by readily available empirical evidence. Standard confirmatory evidence for adaptive accounts includes evidence showing that the different values of a trait (in our case, nonorgasm versus orgasm) are associated with fitness differences, evidence that different rates of orgasm have a genetic basis, and evidence showing that the selective regime postulated was really in effect, for instance, that selection pressures pushing the population in a particular direction really exist (Endler 1986; Lauder 1996). That evolutionary adaptive accounts require this evidence has been standard since Darwin (1964; Lloyd 1983; Kitcher 1985; Rose and Lauder 1996; Sinervo and Basolo 1996). In addition, evidence regarding the description of the trait and how it behaves in various circumstances is relevant to the adequacy of a given account, as is information regarding the phylogenetic distribution of the trait—whether evolutionary relatives show the trait in question or not (see Chapters 3 and 4; Larson and Losos 1996; Novacek 1996).

Under normal circumstances, not all relevant sorts of evidence will be available, but the evolutionary account will be considered well confirmed to the extent that the various types of relevant evidence available or procurable support its various assumptions and explicit commitments (Lloyd 1988/1994, chap. 8; Reznick and Travis 1996; Sinervo and Basolo 1996). That the authors considered in this book accept these very general standards of evidence is suggested by their attempts to describe the trait of orgasm by appealing to the sexology literature, and by their efforts to postulate and support various selective pressures. Moreover, the studies I considered were pursued largely in the context of refereed scientific journals or collections.

In summary, evolutionary accounts of a trait must have scientifically defensible assumptions (as discussed in Chapter 3). If the assumptions of the account are faulty, then the entire evolutionary explanation is threatened.[1] I have documented throughout this book cases in which assumptions that have been contradicted by or unsupported by evidence have been used in evolutionary explanations of female orgasm. Let me summarize the most important of these faulty assumptions here.

## 1. The Assumption That Female Orgasm Is Tied to Reproductive Success, and Thus That It Is an Adaptation

All authors discussed in this book except Symons and Gould make this assumption. It involves an approach to research that assumes that "behavior, even complex social behavior, has evolved and is adaptive" (Barash 1977, p. 8). I discussed the pros and cons of this assumption as applied to female orgasm in Chapter 6. The bottom line, evidentially speaking, is that no evidence has been offered that links female orgasm to either improved fertility or to increased birth rates or reproductive success. If such evidence were available, it would justify the search for an account of how exactly female orgasm does increase reproductive success. Without it, those who take an adaptationist line are relying on a future promise of such evidence being produced. One alternative is to take the available evidence of wide variability in the trait at face value, and to conclude, as Symons does, that female orgasm is not an adaptation at all.[2] As his opponents note, this course of action is unlikely to uncover any links that *might* exist between female orgasm and reproductive success. Thus they argue, assuming an adaptationist stance is a more fruitful research program. The problem we are presented with, though, is that 19 adaptive explanations for female orgasm have been advanced in the absence of the crucial evidence that the trait is an adaptation. Because of this lack of crucial evidence, all such adaptationist explanations should be interpreted as proceeding under an unsupported theoretical assumption, that female orgasm *is* an evolutionary adaptation. The fact that this assumption is part of an entire approach to research in evolutionary biology will be discussed later.

## 2. The Assumption That Female Orgasm Should Be Examined Only as It Appears with Intercourse

This is a tacit assumption. It shows up in researchers' treating female sexuality as if it is equivalent to reproductive sexuality. The reason that this assumption is a problem is the existence of the orgasm/intercourse discrepancy in women, along with the existence of frequent orgasm outside of intercourse. The alternative is to concentrate on the phenomenon of female orgasm per se, as it appears during intercourse, masturbation, and other sexual activities.

Explaining orgasm as it appears only with intercourse may be especially tempting to adaptationists, because such orgasms are thereby tied at least to the possibility of fertilization, since only intercourse is reproductive sex. But given the gaps orgasmic women show in having orgasm with intercourse, concentrating on orgasms with intercourse necessarily distorts the description of the trait to be explained: since there is no a priori reason to link female orgasm with intercourse, female orgasm *itself* (with or without intercourse) should be the target of evolutionary explanations.

It may look as though the recommendation is to treat female orgasm differently than male orgasm, because the existence of male orgasm is explained evolutionarily only as it relates to intercourse. But the cases are indeed different, which justifies their different treatment; male orgasm is known to be necessary for male reproductive success, while the same is not true for women.

### 3. The Assumption That Sexual Intercourse Evokes the Same Response in Men and Women, Namely, Orgasm

Here there is a serious evidential problem: it is well established in the sexology literature that men and women do not necessarily have the same response to intercourse. There are very substantial numbers of women who do not have orgasm with intercourse on any given occasion, and also a substantial minority that *never* do, even though they are orgasmic. Nevertheless, a number of the authors I examined, including Crook, Newton, Campbell, Beach, and Alexander and Noonan, all assume that sexual intercourse unproblematically yields orgasm for both men and women. What is most peculiar about these authors is their ritual citation of the sex literature, despite the fact that the very results cited show exactly the discrepancy they ignore.

Some authors recognize that there is a problem in treating both male and female orgasm as straightforward results of intercourse, but the situation is mischaracterized. Morris, for example, claims that women take longer to reach orgasm during intercourse—a phenomenon that he describes as "strange"—but claims that women generally have orgasm with a long enough period of intercourse. But the evidence from the sexology literature shows that this is not exactly so. Beyond a certain point, women are unlikely to have orgasm, no matter how much longer intercourse proceeds. Hrdy's account is subject to the same problem. Taken in combination with the evidence that men and women take equal amounts of time to reach orgasm during masturbation, this gives additional reason to treat the evolution of female orgasm itself as separate from orgasm with intercourse.

Allen and Lemmon take an entirely different approach to the fact that men and women often have different responses to intercourse.

They acknowledge the orgasm/intercourse discrepancy, but claim that in the past women (with modern genitalia) *did* have frequent and reliable orgasm with intercourse. In doing so they go against all of the sexology literature that says that women's lack of orgasm during intercourse is due to inadequate stimulation because of the physiological mechanics of the situation. Rather than arguing against these findings, which contradict their claim, they simply ignore them.

Finally, I should mention one possible source of confusion regarding the reliability of female orgasm with intercourse. It seems possible that several authors, most prominently Morris and Sherfey, were misled by Masters and Johnson's account of *how* female orgasm occurs during intercourse, when it does, to a conclusion that female orgasm unproblematically occurs during intercourse. Such confusion, while understandable, still ignores the sizable amount of evidence, available at the time, that orgasm does not occur regularly during intercourse for nearly half of women surveyed, and occurred reliably in only 25%.

### 4. The Assumption That Female Sexual Response Is like Male Sexual Response, More Generally

In addition to the assumption that women routinely have orgasm with intercourse the way that men do, we find instances in which females are assumed to respond like males in other ways. For example, Morris claims that in females there is a rapid reversal of the physiological changes associated with sexual excitement. This is precisely what Masters and Johnson documented is not so. It is especially noteworthy that Morris bases much of his description of female sexual response exactly on Masters and Johnson's work, while failing to take their relevant evidence against his claim into account.

Similarly, Gallup and Suarez claim that there is a strong tendency to sleep following female orgasm. But the evidence they offer applies to male orgasm only, and they disregard or are unaware of prominent evidence that females often have tendencies to wakefulness and continued states of arousal after orgasm. In each of these cases, counter-evidence to the authors' claims is ignored, although it was contained in the very sources they cite.

### 5. The Assumption That Female Sexual Interest or Response Is Dictated by Hormones in Prehominids or Early Hominids

There is an assumption shared by a number of authors, including Beach, Morris, Barash, Eibl-Eibesfeldt, Campbell, and Rancour-Laferriere, that the loss of an estrus period in early hominids or prehominids necessitated the evolutionary construction of another motivation for the female to engage in intercourse, namely, female orgasm. These researchers all assume that there was a radical break between hominids and our more distant ancestors, that the hominid females were no longer subject to hormonal dictation of their sexual behavior. The evidential problem with this assumption is that it had already been shown that for nonhuman primates the hormonal cycle did not dictate sexual behavior (see the summary in Dixson 1998).

Another related assumption made by Crook, Eibl-Eibesfeldt, and Barash is that "continuous receptivity" is linked to hominid monogamous pair bonding. Again, relevant evidence against this assumption comes from nonhuman primate data showing some continuous receptivity in species that are not monogamous. In each of these cases, relevant and available evidence from nonhuman primates was ignored.

### 6. The Assumption That Female Nonhuman Primates Do Not Have Orgasm and Therefore That Orgasm Is a Uniquely Human Trait

Morris, Barash, Campbell, and Hamburg all make this assumption, while Beach says that nonhuman primate female orgasms are rare and insignificant. Even Symons, although he reviews some of the evidence for nonhuman female orgasm, declares it unimportant. But there is good evidence that nonhuman female primates show the physiological manifestations of orgasm; some of the evidence comes from hard wiring, some from direct human manipulation, and some from close observation. Before 1970, only anecdotal accounts were available; thus Morris and Campbell should not be faulted for discounting the possibility of nonhuman female orgasm. The other authors, however, are vulnerable to criticism because they ignored or discounted such evidence. In fact, given the development of even better evidence in the 1980s and 1990s, current evaluation of any hypothesis must take this material into account. This turns out to be especially important with regard to recent evaluation of Symons's account, which is actually supported by this later evidence.

There is a persistent peculiarity in the evidence for nonhuman female orgasm: that most of the recorded and observed orgasms occurred in sexual contact with other females, and not during copulation. This leads me to the next problem.

## 7. The Assumption That Evidence for Female Orgasm When One Female Mounts Another Counts as Evidence That Female Orgasm Occurs during Heterosexual Copulation

As we saw in Chapter 5, several of the authors who originally documented cases of female orgasm in the stumptail macaque during female-female encounters went to great lengths to argue that it followed that orgasms also occurred during heterosexual copulation on a regular basis. Goldfoot and colleagues concluded that a minority of stumptail females had orgasm during heterosexual copulation, yet Chevalier-Skolnikoff and Slob and van der Werff ten Bosch argue for the existence of frequent orgasm during intercourse in contradiction to the evidence they offer. They seem to be committed to the prior view that females have the same response to intercourse that the males do.

In all of this work there is a blatant use of two different standards of evidence for female orgasm. Vague, incomplete, or unrepresentative evidence is used to show that most female responses to copulation are orgasmic, in contrast to the patent and straightforward evidence that orgasm occurs during female-female mounting. Slob and van der Werff ten Bosch, for example, present an especially egregious case of special pleading; they introduce a new standard of evidence for female orgasm in order to claim that "female sexual climax may occur during every copulation" (1991, p. 143). And the standard of evidence for female orgasm that Goldfoot and colleagues use is applied rigorously by Slob and van der Werff ten Bosch only in cases of female-female sexual encounters.

This evidential double standard involves a basic belief that the "natural" place for female orgasm is in heterosexual copulation; hence, any evidence for orgasm itself is also ipso facto evidence for orgasm in its "natural" place. This is related to assumption 2, above, that female orgasm should be explained primarily as it relates to intercourse. The problem is that the evidence does not justify this inference. Only direct evidence of female orgasm during intercourse should be used to infer that female nonhuman primates have orgasm during intercourse.

### 8. The Assumption That Female Orgasm Induces a Sucking Motion of the Uterus

Both the Baker and Bellis and the Thornhill and colleagues sperm-competition hypotheses for female orgasm explicitly motivate their views by appealing to evidence for uterine upsuck offered by Fox and colleagues (1970) and Singer (1973). As we saw in Chapter 7, evidence for the upsuck hypothesis is ambiguous at best. While Singer is often cited as offering evidence for the upsuck hypothesis, he offers an interpretation only, and no evidence. Thus the only evidence in favor of the upsuck hypothesis is the two trials done on one woman by Fox and colleagues. This is countered by evidence against upsuck from six subjects in research by Masters and Johnson (1966) and an unknown number of subjects in the study by Grafenberg (1950). Neither the Baker and Bellis nor the Thornhill and colleagues studies offers statistically sound evidence for the phenomenon of upsuck. Moreover, even if it actually exists, the existence of uterine upsuck, by itself, would be insufficient to establish a correlation between either fertility or reproductive success and female orgasm.

## Operative Background Assumptions

Is there something that led these researchers to make such inappropriate assumptions, ones that have serious evidential problems, or is this merely a story about a scientific community working out a research problem over a period of 30-odd years? I think this is patently a story of scientific dysfunction, and I make my case in what follows. I would like to suggest that there are four basic background assumptions in operation that led most investigators either to neglect evidence that was there, to misinterpret evidence, to assume evidence correlating fitness with orgasm was adequate, or to assume that such evidence existed. These assumptions are: adaptationism, androcentrism, procreative focus, and human uniqueness.

### Adaptationism

The most visible background assumption, operating in 20 of the 21 explanations of female orgasm that I have examined, is that of adaptationism. Here, adaptationism involves the presupposition that a trait that evolved served a particular adaptive function for the organism, and that is why it is present in the population. In other words, adaptationists assume that natural selection, rather than other evolutionary forces, directly shaped the trait into its present form, or that natural selection is currently maintaining the trait in the population, and that the explanatory challenge is to discover how. In the case of the female orgasm, the questions become: What selection pressures led to the current form of female orgasm? What is it an adaptation for? What contribution to reproductive success does orgasm make?

But there is a live alternative—Symons's—to the assumption that female orgasm is an adaptation, and thus orgasm's adaptive status cannot be taken for granted without evidence. As emphasized throughout the book, relevant evidence is lacking, which makes adaptationism about female orgasm an unsupported theoretical presupposition. Nevertheless, researchers routinely assume that female orgasm is in fact an adaptation, because they are operating within a theoretical framework in which most interesting traits are assumed to be adaptations. This sets their research agenda, which then becomes the search for the causes of adaptive advantage and the mechanisms of adaptive success.

To fully appreciate what's going on here we need to examine the ideas traveling under the banner of adaptationism in more detail. Let us consider first the research program of adaptationism. The program involves starting an investigation into a trait by working under the operating assumption that female orgasm is an adaptation to some selection pressure. Investigation is continued by measuring the differences in reproductive success owed to different values or manifestations of the trait, locating a genetic basis, or at least heritability, of the variants of the trait, pursuing an engineering analysis of the trait that shows that it is effective for coping with the selection pressure, and documenting that the selection pressure really did exist in the past and continues to do so today (Mayr 1983; Williams 1985; Rose and Lauder 1996; Godfrey-Smith 2001). There are disagreements among researchers concerning which of these or other tasks are most important (see Chapter 6). But all of these approaches begin with the working hypothesis that the trait is, indeed, a special adaptation to help its owner cope with selection pressures.

What happens, though, when the engineering analyses do not work out, or when the trait seems to be inefficient at doing its proposed job? This seems to be a matter of judgment. One alternative is to start over, with the assumption that the trait really is an adaptation, but one where we haven't figured out correctly what it is an adaptation for. This move runs the risk of turning the enterprise into one in which traits are *always* seen as adaptations, even when they may not be. But Williams defends the ad hoc reasons that adaptationists give for their failed predictions, and insists that it has been fruitful—he cites two vivid cases—to resist giving up on the thesis that a specific trait is, in fact, an adaptation (1985, p. 18). The problem is that this seems to lead in the direction of a methodological rule that *all* traits should be considered adaptations at the end of the analysis. It's a matter of not taking no for an answer. But this violates the larger commitment in evolutionary theory to the effect that factors other than natural selection can cause evolutionary change, and that there are important and live alternatives to thinking that all traits are adaptations (Ridley 1996, p. 341). The concern is that if one is doggedly following the thought that a trait is an adaptation, one might miss alternate explanations based on development or correlated characters that are, in some cases, correct. Even to the most careful of adaptationists, it is a judgment call when to countenance alternative explanations.

We can use this issue to sort out various kinds of adaptationists, as I suggested in Chapter 6. One type, which I have called the conservative adaptationists, pursues the various components of adaptationist research and, if an attractive and reasonable nonadaptationist explanation is proposed, is willing to consider it seriously as an alternative. The group that I've called the cavalier adaptationists differ from the conservative adaptationists by avoiding part of adaptationist research, the part in which the different values of a trait are demonstrated to go along with fitness differences. Nearly all of the adaptationist hypotheses discussed in this book are of this type.

The final type of adaptationists I identified was the ardent variety. They describe themselves as ardently pursuing the adaptationist program. This would suggest, correctly, that they attempt to find adaptive explanations for nearly all traits, and that they are generally unwilling to countenance nonadaptive alternative explanations. But this just seems to be a matter of degree; they are less willing than some adaptationists to consider alternative explanations. This formulation of their position is deceptive, however. I shall spell out exactly what the ardent adaptationists' position is in the next section.

For my purposes here, it is enough to note that the cavalier adaptationists avoid a crucial component of the adaptationist research program. In particular, they do not research whether or not the trait of female orgasm is significant for reproductive success; they simply assume that it is. Thus they become oblivious of evidence that it might not be, and are unconcerned about the lack of evidence that this trait makes any difference to reproductive success. This is just the sort of insensitivity to evidence that makes for a pernicious bias, about which more shortly.

There is, however, also a slightly more dangerous element that arises with all sorts of adaptationism, and that is that grossly deficient evidence can be used to justify an adaptationist account. Thus we have Alcock and many others accepting the conclusions of Baker and Bellis and Thornhill and colleagues, evidently without having ever critically examined the papers in question. (There is an unanswered question about how the Baker and Bellis paper ever got published in *Animal Behaviour,* the flagship journal in the field.) The researchers were committed to orgasm's being an adaptation of some kind, and were apparently overeager to find a scenario that supported this view. Thus we have substandard reasoning serving to fill the gaps in an adaptationist account.

## Androcentrism

Androcentrism involves looking at things from an exclusively male point of view, and subsequently neglecting a distinct treatment of a female point of view (Longino 1990, p. 129). With androcentrism, males are taken to be the normal type or the exemplar, while to the extent that they differ from the male type, females are invisible. In the present case androcentrism is implicated in the assumption that female sexuality is like male sexuality in all its essentials, and the result is the concomitant disappearance of an autonomous female sexuality. There are many ways in which female sexuality is, according to the best research, like male sexuality, and in these contexts androcentrism does no harm.[3] But in the cases in which female sexuality differs from that of the male, the background assumption that the male is the normal type tends to obscure important evidence of differences. As we have seen, it turns out that these differences are significant for an accurate evolutionary account.

Androcentrism can be seen in the assumption that intercourse evokes the same response in males and females, namely orgasm. As a generalization, this is clearly false on all the available evidence, yet this assumption appears in numerous adaptive accounts of female orgasm. It amounts to a neglect of an autonomous female sexuality that is distinct from the male pattern. The same is true of the general assumption that females respond in the same ways as males to sexual activity, again an assumption easily shown to be false by the available evidence. In both of these cases, as soon as the evidence from sexological studies is examined, the differences between male and female responses become clear. When androcentrism appears in the present case study, it is striking for its presumption of the equivalence of male and female attributes.

*Procreative Focus*

Procreative focus involves the assumption that all evolutionarily significant sex is procreative sex. This background assumption encourages looking at female sexuality exclusively by focusing on heterosexual intercourse. It is easy to see how this basic approach complements an adaptationist approach: if female orgasm is an adaptation, it must be correlated somehow with increased reproductive success for the female who possesses it. Reproductive success is naturally linked to reproductive sex (heterosexual intercourse); thus it seems natural to think that intercourse is the evolutionarily *important* kind of sex. There are elements of androcentrism and heterosexual bias operating in procreative focus as it applies to female orgasm, because procreative focus concentrates only on the kind of sex that is reliably associated with male reproductive success: intercourse. Thus both adaptationist and androcentric background assumptions contribute to a procreative focus. But heterosexist bias makes procreative focus distinct from a simple adaptationist bias.

Heterosexist bias may be particularly important in considering female orgasm. Both the macaques and the bonobo females have a fair amount of homosexual sex (around half of all female bonobo sexual encounters involve sex with other females). In the bonobos, this is considered to be a form of friendship or coalition building, which can have a profound effect on male behavior (Wrangham and Peterson 1996, pp. 208–210, 227). Thus there may be a fitness advantage in having homosexual sex, and if so, that would make the heterosexist bias a particularly pernicious bias.

Procreative focus promotes the presence of a couple of the evidentially problematic assumptions I have discussed. The assumption that female orgasm should be examined only as it appears with intercourse is a clear example of focusing only on procreative sex.

The result is to treat female sexuality as if it is equivalent to reproductive sexuality, in the face of evidence to the contrary in the form of the orgasm/intercourse discrepancy. In addition, a procreative bias was certainly in play when the primatologists assumed, against their own evidence, that the incidence of female orgasm in nonhuman primates indicates widespread copulatory orgasms in these animals. The authors seem to find it difficult to imagine that female orgasm might be disengaged from the occurrence of reproductive sex. Thus they strain to produce arguments that female orgasm occurs during intercourse, even when it is not observed to do so.

Finally, some sort of procreative background assumption was at work when female orgasm was taken to induce a sucking motion of the uterus, despite the deficiencies of evidence adduced in its support. In this case, an adaptationist background assumption was also involved, since there seemed to be a certain desire to attach female orgasm to something that might affect reproductive success.

*Human Uniqueness*

The final background assumption consists in assuming that human beings are unique in the animal world. There is a strong tendency among some researchers to emphasize the gap between the human lineages and those of our closest relatives. This shows in the assumption of the human uniqueness of various traits, including aspects of female sexuality. For example, take the assumption that female sexual interest or response is dictated by hormones in prehominids or early hominids. Here, both our ancestors and nonhuman primates are taken to be significantly different from modern human beings, whose sexual activity may be affected but is not dictated by hormonal activity. There is a sharp line drawn between full-fledged hominids and others, based on the supposed differences in hormonal control of female sexuality. But the evidence indicates that the situation is much more fluid than this, and that, in fact, some nonhuman primates may be more like modern human beings in the lack of hormonal dictation of their sexual behavior.

## The Deepest Problems

The background assumption of human uniqueness has tended recently to be out of play, and will not be discussed further. Procreative focus involves both adaptationism and androcentrism, and I shall proceed by exploring these latter two background assumptions. According to my analysis, some of the evidential problems in proposing and evaluating evolutionary explanations of female orgasm arise out of a background assumption of androcentrism. (Note that I am assuming throughout that commitment to a background assumption may not be the result of conscious deliberation.) In these cases, which involve not treating female sexuality as autonomous from either male sexuality or reproduction, it can seem sensible to focus on reproductive sexuality as the basis for understanding female sexuality. After all, the intuition goes, what is sexuality for beyond the fundamental function of reproduction? There is much intuitive support for the notion that evolutionary changes in female sexuality *must* be related to increases in the reproductive fitness of those females exhibiting the changes. The problem is that the evolutionary explanations created using this strong intuition have failed to track what is independently known about female sexuality. Female sexuality, particularly female orgasm, does not seem to follow this line of reasoning when taken from the intuitive standpoint that female sexuality evolved in tight concert with reproductive success. Once we add the sexology literature as a basis of evidence, it is easily seen that crucial aspects of female orgasm are being omitted from the androcentric evolutionary pictures. Such evidence demands an account of female orgasm that is autonomous and distinct from both male sexuality and reproductive sexuality.

Perversely, the only available explanation for female orgasm that does make it independent of both male sexuality and reproductive sexuality has been repudiated for being androcentric. Again, it is easy to see how this happened. The byproduct account appears, at first glance, to preclude the possibility of an autonomous female sexuality because of its emphasis on the common embryological origins of human males and females and on the selection pressures on male orgasm. But interestingly, the byproduct view severs the links between adult female sexuality and reproductive sex. It fully acknowledges and in fact accounts for the very features of female orgasm that the androcentric accounts ignore or attempt to explain away. Thus in my view this is the evolutionary account with the closest ties to the feminist value of separating definitions of women—including women's sexuality—from women's reproductive functions. It is also an evolutionary view that avoids the errors of assuming that female response is like male response both during intercourse and more generally. Thus it avoids all of the evidential problems produced by androcentric values that appear in many of the other available explanations.

It seems that, ultimately, the vision of female orgasm in the byproduct view is considered androcentric precisely because a trait must be an adaptation itself in order to be considered genuinely culturally valuable. But this does not follow. Such a requirement would require accepting an extreme sort of adaptationism, and feminists have no independent reason—including any political reason—for buying the equation between adaptation and cultural significance. Feminists do maintain an interest, however, in continuing to fight for definitions of women that are not based on their reproductive roles. Hence, I conclude that in the feminist objections to the byproduct theory a nonfeminist value placed on adaptation has superseded, without a good reason, a legitimate feminist value placed on separating women from their definition in terms of reproductive role.

Let me be very clear about this. I am not claiming that I prefer the byproduct hypothesis on the basis of any social value, nor am I suggesting that anyone else do so. Nevertheless, Longino has emphasized that social values may, in particular cases, "have a positive role in grounding criticism of background assumptions and in fostering the development of empirical investigation in directions it would not otherwise go" (2002, p. 51; Campbell 1998, 2001). Hence, *if*—and I must emphasize the "if" here—social values do play a role in the full evaluation of the hypotheses and background assumptions under question, surely feminist values would be more in tune with the byproduct account, given its separation of female orgasm from the function of reproduction. However, I prefer the byproduct account primarily because it is supported by overwhelmingly better evidence (modulo later analysis in this chapter).

The other major source of evidential problems in the evolutionary explanations I have examined is the background assumption of adaptationism. Thus two overarching and tightly interwoven kinds of background assumption have played the major roles in constructing and evaluating evolutionary explanations of female orgasm: androcentrism and adaptationism. The androcentric themes are intimately entwined with the adaptationist ones. Once we are looking for an explanation that ties female orgasm to reproductive success, we are virtually inexorably driven into procreative and androcentric biases. The one explicitly feminist explanation, Hrdy's, still contains a procreative focus that leads her, it seems, into a misunderstanding of the sexology evidence about female orgasm. Of the other early accounts, only a few, such as Alcock's, Rancour-Laferriere's, and Bernds and Barash's, escape overt androcentrism, and these fail to provide the necessary supporting evidence for their hypotheses. In all of the other cases, adaptationism and androcentrism push the explanations toward an exclusive focus on female orgasm with intercourse, thus bypassing the very data that suggest that an explanation of female orgasm independent from intercourse may serve as the best evolutionary approach. The result is a shared set of androcentric and adaptationist background assumptions, which lead, I have argued, to the sorts of unacceptable treatment of evidence described in this book.

So, where are we now? I first reviewed assumptions in the evolutionary explanations that conflict with one or another body of evidence relevant to the evolution of female orgasm. I have now tried to isolate a set of overarching background assumptions that could be used to motivate, defend, or justify those faulty assumptions. I conclude that because they neglect or mistreat empirical evidence these four background assumptions are operating as pernicious or damaging biases in the pursuit of the explanation of female orgasm. A traditional approach to bias, as exemplified by the logical positivists, for example, poses bias as a systematic deviation from the normal methods for ensuring the objectivity of scientific results (Carnap 1962; Laudan 1984; Richardson 1984; Geertz 1990; Haack 1993; Sober 1993; Gross and Levitt 1994; Reichenbach 1996). Under this view, systematic deviations from objective rules are explained by thinking that scientific reasoning includes considerations that are irrelevant to the truth of theories or claims. This is hypothesized to occur especially because theories accord with preconceptions and special interests of human investigators.

One could argue that androcentrism embodies exactly the sort of bias rejected on this traditional account. But such a view turns a blind eye to how such biases are detected and corrected. Social values are involved in applying a corrective feminist view, which later helps eliminate androcentric bias. Thus it seems as if using social values in evaluating science is a good thing that contributes to the objectivity of scientific findings. But this conclusion is usually denied by those using the traditional approach, who separate the process by which bias is uncovered from the end result of reducing bias (Gross and Levitt 1994). Thus androcentrism could be accounted for by using this traditional approach to bias.

But the traditional approach would not be adequate for the present case study, in which adaptationism also plays a biasing role. Adaptationism does not embody the sort of social bias that is rejected on the traditional account. It is part of a legitimate research approach in evolutionary biology, and does not seem to incorporate social bias in the way that androcentrism does (Rose and Lauder 1996). In fact, the evidential standards for demonstrating that a trait is an adaptation are well worked out and quite strict (Lauder 1996; Sinervo and Basolo 1996). The fact that adaptationism appears to be a problem in this case study suggests that the practices within evolutionary biology bear a closer look. This involves a contextual approach to science, a set of views developed to deal with the shortcomings of the traditional account.

So far, it looks as if background assumptions are themselves a bad thing for science. But philosophers of science have long argued that background assumptions are inevitable in science, and necessary to it.[4] If this is right, then the real question becomes: How are we to adjudicate between background assumptions that are beneficial for scientific practice and those that are not?

No argument in this book is intended to establish that adaptationism and even androcentrism always lead by themselves to inadequate science. It is likely that much good science could be done with some of these background assumptions in the form of social or theoretical values in place.[5] What I am considering is rather whether adaptationism and androcentrism have had a destructive effect on the discussion of the evolution of female orgasm, in particular. Judgments about whether particular background assumptions are damaging must be made on a case-by-case basis. What I have presented is a particular case in which both androcentrism and adaptationism have led to inadequate science judged at the most basic level of evidential support.

## Science and Background Assumptions

The operating assumption so far in this chapter has been that there is some neutral set of data to which a theory must be accountable. But this position is philosophically naive. Questions of adequacy to the data involve issues concerning *which* data a theory ought to be accountable to, which in turn involve a much larger vision of both how science actually works and how it should work (Longino 1995, p. 394). For the rest of this chapter I shall make use of Helen Longino's well-developed framework for how science does and ought to work. According to Longino, "commitment to one or another model [theory] is strongly influenced by values or other contextual features. The models themselves determine the relevance and interpretation of data" (1990, p. 189). In this particular case, I would argue that the sexology literature is treated as a source of data to which all parties appeal in the debate about the evolution of female orgasm. Nevertheless, we need to explore further the relations between hypothesis and data. How, exactly, do background assumptions influence the choice of theories, and what effect does this have on the treatment of evidence? Can we glean any answers to such questions from the present case study?

According to Longino, the subjects of investigation are constructed, not simply given by nature. Any inquiry, she writes, "must characterize its subject matter at the outset in ways that make certain kinds of explanation appropriate and others inappropriate. This characterization occurs in the very framing of questions" (1990, p. 98). In the present case, inquiry into the evolutionary origins of female orgasm requires an evolutionary answer. But the situation is more complicated than that. Specification of the subject of investigation can depend in turn on the needs and interests of the questioners (Anderson 1995, pp. 44–46). For example, those following an adaptationist approach may frame the question: What is the evolutionary function of female orgasm? Or: What selection pressures led to the current form of female orgasm? Others, not following an adaptationist program, may ask a broader question: What accounts for the evolution of female orgasm? Thus background assumptions and values can play a role in specifying the very object of inquiry.

Longino offers a careful analysis of the role of background assumptions in evaluating evidence, and sees their role in science as inevitable. The most fundamental aspect of the relation between data and hypothesis is that it is flexible and changeable. The basic problem is that states of affairs (data and observations) do not tell us what they are evidence for. The same data can be thought of as evidence for two separate hypotheses, or for one hypothesis for two different reasons. This is a well-established result in philosophy of science, the underdetermination of hypotheses by evidence.[6] In order to link particular data to a particular hypothesis, additional assumptions need to be brought in concerning the evidential relation between facts and hypotheses. Thus the evidential relevance of hypotheses to observations or experiments is a function of the background assumptions that help drive the inference. Thus, technically, we have only "data" until background assumptions make it "evidence" for or against a particular hypothesis.

Background assumptions can be of many sorts. Oftentimes they are simply assumptions about what there is in the world; sometimes they are simple inductive rules. But often they are more substantive than this, and they may bring in social and individual values, interests, and subjective preferences (Longino 1990, p. 48). Such preferences and values can be seen as a source of the biases that appear in the proposal of hypotheses and in their evaluation. I have examined some of the background assumptions that serve as sources of pernicious bias in evolutionary explanations of female orgasm, including adaptationism, androcentrism, procreative focus, and a focus on human uniqueness.

I want to approach my case study by couching the biases I've outlined in terms of their being background assumptions. I am still left, however, with the difficult task of discriminating background assumptions that are negative for science from those that are beneficial or neutral. Intuitively, the result here should be that androcentrism, adaptationism, procreative focus, and human uniqueness serve as pernicious background assumptions or biases, in these particular cases of scientific reasoning. Intuitively, an illustration of a beneficial background assumption will help. One higher-level background assumption important to this research is the assumption that not every biological character is adaptive—that there exist alternative evolutionary explanations available, such as evolutionary developmental accounts or accounts that cite correlations of growth (Ridley 1996; Griffiths et al. 2002). This is a background assumption of evolutionary biology quite generally, and it is clearly relevant to the case of female orgasm. What we have to watch, though, is whether this assumption is actually engaged when researchers are arguing about whether female orgasm is an adaptation or not. There is an important difference between paying lip service to the view that there could be acceptable alternative explanations for a trait, and actually using this assumption when reasoning about a specific trait. As I shall argue below, there are adaptationists in the female orgasm debate who seem to find any explanation under which female orgasm is not considered an adaptation unacceptable, despite the fact that nonadaptive traits are one important possibility in evolutionary theory. This raises the question just mentioned regarding whether anything more than lip service is being paid to the foundational assumption from evolutionary biology that alternative, nonadaptive explanations are part of the toolkit of evolutionary theory.

One approach to picking beneficial assumptions from those that are not has been offered by Elizabeth Anderson, in an expansion of Longino's work. Alvin Goldman and Philip Kitcher treat bias as a "kind of self-interested motivation that certain things be true rather than others" (Kitcher 1993; Goldman 1995; Longino 2002, p. 166; see the citations to other traditional approaches). Longino's and Anderson's approach, in contrast, expands the role of biases to include background assumptions, which can be treated as potentially negative (but also potentially positive) biases. According to Anderson, pernicious bias is a failure of impartiality. Miriam Solomon treats bias as a descriptive concept. On her psychologistic view, a bias is anything that focuses attention or inclines belief in a certain direction. Biases, on Solomon's view, are productive; therefore there is no drive to eliminate bias (1995; 2001). The position taken in this book is normative rather than descriptive, and finds that certain biases can have a negative impact on scientific inquiry (see the discussion in Longino 2002, pp. 165–167; Campbell 1998). We can take Anderson to be defining when a background assumption has a damaging effect on the science.

When it comes to scientific inquiry, impartiality must be related to the specific goals of a particular inquiry, to the scientific question being asked: in my case, by what means did women evolve to be orgasmic? Like Longino, Anderson notes: "all inquiry begins with a question. Questions direct inquiry by defining what is to count as a significant fact and what is a complete or adequate account of a phenomenon" (1995, p. 42). She continues: "What counts as a significant truth is any truth that bears on the answer to the question being posed. The whole truth consists of all the truths that bear on the answer, or, more feasibly, it consists of a representative enough sample of such truths that the addition of the rest would not make the answer turn out differently" (1995, pp. 39–40).

According to Anderson, impartiality or lack of pernicious bias demands attention to all facets of available empirical data, including those that weigh background assumptions.[7] In particular, impartiality requires taking into account data that support alternative hypotheses, or that are inconsistent with one's assumptions. The philosophers Paul Feyerabend, John Stuart Mill, and Helen Longino all support the importance of alternative hypotheses. Feyerabend assumes, like Longino, that theoretical assumptions are ineliminable, and he endorses the development and pursuit of alternative hypotheses for the sake of increasing the number of facts with which we can evaluate them (Longino 2002, p. 129; see the discussion of Feyerabend and Mill in Lloyd 1997). Anderson's view implies that both relevance and representativeness of data are legitimate criteria on the basis of which to choose theories.

What we have, in many of the explanations for female orgasm reviewed in this book, is a failure of impartiality. Most frequently, data that bear on the question being posed have been either ignored or misrepresented in the suggested accounts of the evolution of female orgasm. I have already reviewed some of the detailed instances of pernicious bias—failure of impartiality—in the first part of this chapter. In particular, relevant data from sexology and primatology were ignored. Thus this is partial science according to Anderson, and can be critiqued for being biased in a destructive way. There is a way in which Anderson's approach is reminiscent of a traditional approach to bias, and this is a strength. She appeals to a methodological standard—a version of the requirement of total evidence—and rejects deviations from that standard.

The discussion has so far been concerned with how to evaluate particular background assumptions in relation to a specific set of evidence. But I now turn to criteria of the capacity of a community to produce objective scientific inquiry that is based on how the community as a whole interacts. This community approach has the advantage of moving the focus away from individual idiosyncrasy, and instead focuses on community-wide standards of investigation. This will help formulate what has gone wrong within the community investigating the evolution of female orgasm.

According to Longino, community-level criteria can be used to discriminate among the products of scientific communities and, especially, to neutralize personal idiosyncrasy and challenge background assumptions.[8] This approach requires that we consider all the people working on the evolution of female orgasm as a community of scientists, all aimed at solving a basic problem. The key to the process of arriving at a community with the capacity to produce objective scientific findings is the fulfillment of four criteria, which are seen as necessary to achieve a level of discussion where scientific mistakes that are made may be corrected through the critical interaction of the members of the community (1993b, pp. 112–113).

Longino's first criterion for objective inquiry is that there must be "publicly recognized forums for the criticism of evidence, of methods, and of assumptions of reasoning" (1993b, p. 112). This requirement would seem to have been fulfilled adequately for many of the explanations I have examined. Perhaps the best example of this is the presentation of Symons's byproduct hypothesis in *Behavioral and Brain Sciences,* with its batch of peer commentaries published in the same issue (1980a, 1980b). The journal *Animal Behaviour* has also served as a site for the publication of much critical and analytical discussion of adaptationism and female orgasm.

Longino's second criterion for objective knowledge production is that the community must change its beliefs and theories over time in response to the critical discourse among community members (1993b, p. 112). Again, this requirement would seem to be at least partly fulfilled in the female orgasm case, given the changes in favored explanations over the 37 years of research I have examined. Symons (1979) and Hrdy (1981) gave early and harsh criticisms of the early pair-bond approaches, which I find relatively undefended after these criticisms. Hrdy's own account, in turn, was criticized, and she responded by admitting that there were gaps in the evidence for her theory, and by emphasizing that it is not an account of current adaptation. Symons's account too was criticized, but was later defended by Gould, who also responded to the subsequent attack by Alcock. In contrast, although Baker and Bellis's account was sharply criticized by Short and by Dixson, there seems to have been no response to that criticism. In particular, Short's claim that sperm competition does not occur in human beings—coming as it does from a founder of the study of sperm competition in primates—appears fatal to both the Baker and Bellis and Thornhill and colleagues accounts, but it has not been rebutted. And, in fact, Baker and Bellis's account is widely accepted by both scientists and the lay public.

Longino's third criterion for objectivity is an especially demanding one: "There must be publicly recognized standards by reference to which theories, hypotheses, and observational practices are evaluated and by appeal to which criticism is made relevant to the goals of the inquiring community" (1993b, p. 112). The general family of standards can include "substantive content, criteria of evidence and reasoning, and methods of investigation" (2002, p. 148).

Here, there seem to be problems in my case study. Although the theorists advancing evolutionary hypotheses fully recognized the available sexology literature as providing evidence relevant to the theories in question, they used that evidence very selectively and sometimes misrepresented it. In addition, many theorists assumed that adaptationism was a publicly recognized and accepted background assumption, while others did not. Thus while there was, indeed, some response to criticism, as discussed above, it was far from complete and thoroughgoing. This situation will be discussed in more detail in a moment.

The final criterion for effective inquiry is that communities must be characterized by tempered equality of intellectual authority. "What consensus exists must not be the result of the exercise of political or economic power or of the exclusion of dissenting perspectives; it must be the result of critical dialogue in which all relevant perspectives are represented" (1993b, pp. 112–113). Since there is, in fact, no deep consensus about the evolution of female orgasm, it might seem that this criterion does not apply to the female orgasm debate. Nevertheless, there has been a fairly clear dismissal of the byproduct view on nonevidential grounds. For programmatic reasons favoring adaptationist explanations, the dissenting nonadaptationist view has been ridiculed and dismissed. Thus this view, although relevant, has failed to be fully represented in recent discussions. To the extent that there is now a consensus favoring the sperm-competition views that cannot, if my analysis is correct, be defended on evidential grounds, there might be a failure of equality of intellectual authority at work. In particular, it seems that the authority of an anthropologist (Symons) and a paleontologist (Gould) might not carry weight with those working in contemporary animal behavior, who are very much against the byproduct account.

The effectiveness of a given community of inquirers in producing objective scientific inquiry can be evaluated according to how well it fulfills the criteria listed above. The more completely a community fulfills the criteria, the more objective that community is. This standard is based on how well a scientific group incorporates the elements needed to have ongoing, critical debate.

As mentioned above in the discussion of the third criterion for objectivity of a scientific community, Longino requires that the community engage in critiques of its own background assumptions using publicly recognized standards. Anderson gives us one basis on which to evaluate background assumptions, in terms of their impartiality, which is a relation to a set of data. Once we examine the background assumptions of androcentrism, human uniqueness, and procreative focus, we can see that they are implicated in partial treatments of the data, in which relevant data are ignored. The same is true for adaptationism, though the situation here is more complicated. Thus Anderson's criteria can help us evaluate whether specific background assumptions are implicated in inadequate science. The standards for the background assumptions in the female orgasm story do seem to be shared for the most part, except for the case of ardent adaptationists.

I should note that even though evidence can be used to challenge background assumptions, there are yet further background assumptions required to link the challenged background assumption to this evidence. In general, background assumptions are challenged with relation to a particular set of other background assumptions and the public standards of a community (Longino 2002, pp. 127–129). The real benefit to the community lies in the effective critical interactions that are generated in challenging background assumptions. Following Mill and Feyerabend, Longino writes, "Criticism not only spurs evaluation and reevaluation of hypotheses, but also leads to better appreciation of their grounds and their consequences" (2002, p. 129).

One reading recent summaries of the situation with regard to female orgasm would come away with the impression that there is really only one live account being pursued, and that is the sperm-competition account (Buss 1994; Fausto-Sterling et al. 1997; Singh et al. 1998; Daly and Wilson 1999; Jolly 2001; Rodgers 2001; Barrett et al. 2002; Singh et al. 2002). Yet few, with the exception of Short and Montgomerie and Bullock, seem to be examining the actual merits of this account. In particular, no one seems to have noticed the most fundamental flaws of all, which are that there is at best equivocal evidence that uterine upsuck actually occurs, and that there are fatal statistical problems with the Baker and Bellis data and interpretation. This is, indeed, a failure of basic critical examination, and seems to be affected by theoretical commitments to the idea that female orgasm must be an adaptation, at least among some of the theory's supporters.[9]

Longino's criteria for community objectivity are quite general, and determining exactly how her criteria can be applied to the female orgasm case is nontrivial. Under my expanded analysis of Longino's third criterion, there are three major components of the evaluative standards that are not completely agreed upon by the practitioners I am considering. On my reading, evaluative standards include standards of which questions to ask; standards of which evidence is relevant, and appropriately established; and standards of which kind of explanation are appropriate. Let us see how each of these types of evaluative standard differs between the adaptationists and those who support the byproduct view.

In my case study, the object of inquiry seems to be defined differently for the two groups, as discussed above. That is, for adaptationists the question is: What selection pressures led to the adaptation of female orgasm and what is its contribution to reproductive success? While for the byproduct advocates, the question is: How did the trait of female orgasm appear and how is it maintained in the population? Hence, we have two quite distinct sets of questions driving the research efforts of adaptationists and byproduct theorists.

Regarding standards of evidence, there has been plenty of critical dialogue concerning the merits of the byproduct view and of various adaptationist accounts (see Chapter 6). Nevertheless, the two groups seem to be adopting different standards of evidence. I ended Chapter 6 by noting that the adaptationists believed that recent data from Baker and Bellis established the elusive claim that there is a connection between female orgasm and reproductive success. As I showed in Chapter 7, however, using completely customary standards of statistical evidence, this conclusion is simply unwarranted. Thus we are back to the long prevailing state, one with a lack of evidence supporting the claim that female orgasm is a biological adaptation, that its presence is associated in some way with reproductive success. This is the basis of the claim, by those who support the byproduct analysis, that there is no evidence that female orgasm is an adaptation, and thus that adaptationist explanations of the trait are unnecessary. A closer look at what is at issue between the adaptationists and the byproduct advocates makes it clearer that these two groups are using slightly different standards in evaluating their hypotheses—although they agree on the standards of evidence that I use in the first section of this chapter, to implicate the various assumptions.[10] In addition, they endorse different standards of explanation; in particular, the adaptationists rule out nonadaptationist explanations of female orgasm a priori.

As I noted in Chapter 6, ardent adaptationists and those supporting the byproduct view use different standards of evidence. In particular, the two groups treat both positive evidence and lack of evidence differently. In the case where there is a lack of evidence for a connection between female orgasm and reproductive success, the ardent adaptationists assert that there either is or must be such evidence, while the supporters of the byproduct view take the lack of such evidence at face value. In addition, ardent adaptationists take a very negative view of the supporting evidence for the byproduct approach, considering it either invisible or useless. They see the byproduct view as a sort of "giving up" on evolutionary explanation (that is, on adaptive explanation). Evidence supporting the byproduct view is never discussed in detail; however, the ardent adaptationists do accept a very limited version of the byproduct view. Specifically, they accept as a fact that female and male orgasmic tissues arise from the same embryological roots. But they assert that this has nothing to do with why the trait of orgasm appears in the population of females over evolutionary time or why the trait is maintained today.

It is here we run into the different standards of explanation that are used by the ardent adaptationists and the byproduct view supporters. According to Alcock (1998), byproduct explanations can never be evolutionary explanations at all, because they highlight only the proximate (developmental or physiological) cause of the trait. Under this view, in order to be an evolutionary explanation at all, the account must give an adaptive explanation for female orgasm. Clearly, this is dogmatic. A related view is advanced by Sherman (1988), in which byproduct explanations are seen as concerning only how female orgasm came to be in the population, and not whether or not it is related to reproductive success, which he says is the heart of the matter for Alcock. Here, Sherman makes the concrete assumption, unsupported by any cited evidence, that female orgasm is correlated with reproductive success.

A further example of the different standards of explanation that are in place is the pragmatic argument put forth by the ardent adaptationists. Here, the argument is that if we accept a byproduct account of female orgasm, then that will prematurely stop the search for a viable adaptive account. This is deemed to be a bad thing, especially under the assumption that the trait is, in fact, an adaptation. Thus, given that an adaptive account may be just around the corner, we must not accept the byproduct account. The ardent adaptationists' mistaken view that there is no supporting evidence for the byproduct account comes in handy here. They treat the byproduct account as a kind of null result, akin to scientific surrender. It is clear that the ardent adaptationists do not see a byproduct explanation as being on equal footing with an adaptive explanation. Under this argument, however, the situation is worse than that: it seems that no byproduct explanation should *ever* be accepted.

One very important background assumption is in play in these discussions about what kind of explanation would be adequate. Remember that one of the significant background assumptions for evolutionary biology is that there are a variety of mechanisms that can account for the evolution of an organism or a trait (Ridley 1996, p. 341; Griffiths et al. 2002). Selection, correlations of growth, drift, and developmental noise are various causes that can account for the manifestation of a particular trait. Lip service to this basic assumption is paid by all. However, in this case it seems that this assumption is being overridden; an adaptive explanation—one in which *only* selection is seen as an effective cause—is seen as the only really acceptable explanation for female orgasm. Other accounts are not seen as adequate. Thus here we've got the denial of a basic background assumption of evolutionary biology. I suggest that this puts the ardent adaptationists on a different page from the cavalier adaptationists, who are not committed to the denigration of alternative accounts of a trait. The cavalier adaptationists may share the basic standards of evidence and explanation with the supporters of the byproduct view, while the ardent adaptationists seem to differ from both groups.

In other words, it seems that we've got a subgroup of adaptationists, the ardent adaptationists, who hold both standards of evidence and standards of explanation different from those of both more mainstream adaptationists and the supporters of the byproduct view. This makes fulfillment of Longino's third criterion—that researchers accept shared standards of evaluation—unattainable in the case of the ardent adaptationists. To the extent that ardent adaptationists disagree with the standards of evolutionary biology more generally, the reasoning processes involving their claims are going to be unacceptable to other evolutionary biologists. This makes it difficult for the types of challenges and responses that Longino says are required for the kind of community that is the best at producing objective inquiry.

If the two sides really do show a lack of agreement about standards of evaluation, then we have a violation of criterion 3. Let us return to this third criterion, the demand for publicly recognized standards by reference to which hypotheses and observations are evaluated. According to Longino, there are two general types of criticism to which appeal can be made in this context. The first is "evidential" criticism, which "questions the degree to which a given hypothesis is supported by the evidence adduced for it, questions the accuracy, extent, and conditions of performance of the experiment and observations serving as evidence, and questions their analysis and reporting" (1990, p. 71). This is the sort of criticism that I brought to bear on the Baker and Bellis and Thornhill and colleagues hypotheses and experiments. It covers the most basic types of empirical adequacy of hypotheses, and can be supported by appeal to widely shared standards, such as statistical practices.

A second type of criticism, "conceptual" criticism, cuts much more deeply into the assumptions underlying a theory. It can involve criticism of the conceptual soundness of a hypothesis, examination of the consistency of the hypothesis with accepted theory, and questions regarding the relevance of the evidence presented in support of a hypothesis (1990, p. 72; see Rooney 1992; Nelson and Nelson 1994). This type of criticism amounts to questioning the background assumptions in light of which the hypothesis is proposed and evaluated, and it is crucial for the achievement of objective inquiry. The present case, with the different treatments of the lack of evidence for orgasm being an adaptation, is ripe for conceptual criticism. I have illustrated what conceptual criticism of this case would look like, especially with regard to byproduct versus other evolutionary accounts of traits, in this chapter. Neither side accepts the other's standards for when evidence is crucially required for the endorsement of a hypothesis. They are at an impasse. Nevertheless, any adaptationists coming in and attempting to show a correlation between female orgasm and reproductive success could overcome the impasse. Then, at least, the nonardent adaptationists and the byproduct supporters would be on the same page regarding basic evidence necessary for an adaptive explanation.

In fact, note that with all of the cavalier adaptationist accounts, they do not take a dogmatic stand like the ardent adaptationists, they simply make an assumption—in some cases, it looks unthinking—that female orgasm is an adaptation. Thus it seems to be unfair to group all of the adaptationists together. With most of them, there is simply a lack of evidence for a crucial assumption of their explanations, while in the case of the ardent adaptationists, there seem to be different standards of evidence and explanation operating altogether. In other words, it seems that the conservative and cavalier adaptationists and the byproduct supporters may well be operating within a single set of standards for the evaluation of a theory, while the addition of the ardent adaptationists breaks up the community unanimity of evaluative standards.

In summary, then, Longino gives an account of community objectivity in which background assumptions play a central role. As such, Longino's account, in combination with my expansion of it, is very helpful in pinpointing exactly where the ardent adaptationists part ways with others in the community of evolutionary biologists, at least on the issue of the evolution of female orgasm. Thus we have diagnoses of what went wrong in the science I discussed. In the case of androcentrism, this background assumption ought to be rejected because it leads to missing a relevant chunk of evidence about autonomous female sexuality. And ardent adaptationism should be rejected for conflicting with a basic tenet of evolutionary biology, while cavalier adaptationism leads to a disregard of crucial evidence.

## Conclusions

I have pinpointed a set of problematic background assumptions that appear in various explanations for the evolution of female orgasm. Background assumptions can be challenged and rejected in the course of the transformative criticism necessary for the production of scientific knowledge. In this case, I would propose jettisoning all four of the background assumptions I have considered—ardent and cavalier adaptationism, androcentrism, procreative focus, and the focus on human uniqueness—since they have all been implicated in badly reasoned or badly supported evolutionary explanations. There is a direct tie between making these background assumptions and making the specific evidential errors that I have detailed in this book. This is not a claim about what necessarily must be the case with these overarching assumptions; rather, it is about what actually has been the case when it comes to explanations of female orgasm.

In summary, Longino's analysis of community objectivity in science commits us to the idea that background assumptions are always in operation in science. As noted before, this approach is in contrast to accounts of bias that view it as a personal preference for a particular outcome. Anderson's discussion of impartiality gives us a way to discriminate between harmful biases and beneficial ones. In particular, harmful biases lead to partiality in the treatment of data, which is exactly what I find in nearly all of the evolutionary explanations I examined. I can also use Longino's analysis of objective inquiry to help determine what went wrong in the scientific investigation of the evolution of female orgasm, especially with regard to the ardent adaptationists. Specifically, these researchers seem to fail the requirement of having a shared set of standards with which to evaluate hypotheses. In contrast, I have tried to use a minimal and widely shared set of basic empirical standards on which to judge the various accounts.

I take myself to have demonstrated the following: first, that there are serious evidential problems with all but one of the available evolutionary explanations for female orgasm; second, that certain background assumptions, especially adaptationism and androcentrism, are centrally implicated in the scientific failures so far, and thus both theoretically and socially motivated factors are intertwined in producing the inadequate science I have examined in this book; and third, that there is much fruitful research that could be done to help advance the state of knowledge. In conclusion, I would like to emphasize the open research questions that are apparent when this material is examined critically. One line of research involves pursuing an implication of the byproduct view. Cross-species comparisons could be done to determine whether those species having highly sexual males also have females who are capable of having orgasm. If this is true, it would lend support to the byproduct view. In addition, cross-species studies of female anatomy, especially those involving the placement of the clitoris, would help clarify the question of whether direct clitoral stimulation is obtained during intercourse for any of our close relatives. This is a line of research strongly supported by Hrdy. It would also be extremely useful to have data collected regarding the frequency of women who have orgasm from assisted as opposed to unassisted intercourse, and there is also a dire need for further, accurate cross-cultural evidence regarding the frequency and circumstances of female orgasm. Research regarding a possible correlation between orgasm and pregnancy timing might also contribute to an adaptive account. Finally, pursuit of the effects of oxytocin on uterine contraction is needed. Most of the oxytocin research is currently being done within the context of fertility studies. If there is, indeed, a correlation between female orgasm and fertility, this is the most basic evidence necessary for any adaptationist account of female orgasm, and it is still missing.

The history of evolutionary explanations of female orgasm is a history of missteps, misuse of evidence, and missed references. The case is still open, and it is ripe for some good scientific work.

NOTES

BIBLIOGRAPHY

ACKNOWLEDGMENTS

INDEX

# Notes

## 1. Introduction

1. The C allele of hemoglobin also provides malarial resistance, but it is not lethal in homozygotes, and hence represents a different set of selection pressures. The hitch is that it is recessive for malarial resistance, and hence requires two copies of the gene in order to be effective.

## 2. The Basics of Female Orgasm

1. Bancroft 1989, pp. 84–86; Mah and Binik 2001, pp. 839–840. In laboratory studies and self-reports, about 66% of women subjects reported an erotically sensitive area on the anterior vaginal wall, an area that can be stimulated to the point of orgasm. (See Addiego et al. 1981; Perry and Whipple 1982; Goldberg et al. 1983.) There is some evidence that stimulation of this zone induces a urethral ejaculate at orgasm. The ejaculate itself may have characteristics similar to the fluid from the prostate in males (Zaviacic et al. 1984; Altshuler 1986; Zaviacic et al. 1988; Darling, Davidson, and Conway-Welch 1990; Zaviacic and Whipple 1993; see the summary in Mah and Binik 2001, pp. 839–840).

2. Mah and Binik 2001, pp. 838–839. See the discussion in Bancroft 1989, pp. 82–83. The chief proponent of this view is Singer (1973), although other possible supporters include Komisaruk and Whipple (1995). Conflicting evidence comes from Kinsey et al. (1953); Alzate (1985a), and Hoch (1980). Retrospective and prospective studies are

in conflict about the orgasmic importance of the cervix (Mah and Binik 2001, p. 839). One study of women who had had their clitorises removed found that most of the women could have orgasm with intercourse (Lightfoot-Klein, 1989). Lightfoot-Klein gives some examples of descriptions of those orgasms that I find unconvincing. If it is true that some of these women experience orgasm, it would be due to the deep stimulation of tissues surrounding the vagina and the uterus.

3. See Bohlen et al. (1982) for an alternate view of the role of these tissues in orgasm.

4. See Hite 1976, 1981; Alzate 1985a; Bancroft 1989; Schiavi and Segraves 1995; Mah and Binik 2001.

5. Robinson (1976) has criticized the Masters and Johnson model on the grounds that the excitement and plateau phases are not identifiably distinct.

6. When multiple orgasms occur they do not ejaculate until the last orgasm (Kinsey et al. 1948, pp. 158–159; Robbins and Jensen 1977). Kinsey and colleagues reported the occurrence of multiple orgasms in approximately 55% of preadolescent males, and in 3% of men over the age of 30. Among younger adult males, Kinsey found 8–15% capable of multiple orgasms (Kinsey et al. 1948, p. 376).

7. Multiple orgasms in women are documented in Masters and Johnson (1966), Hite (1976), Bohlen et al. (1982), Amberson and Hoon (1985), and Darling et al. (1991).

8. This calculation treats Gebhard's 59% subsamples as separate data points. The other subjects in one of his studies reported a 35–38% orgasm rate, so my calculations of mean and median are probably high.

9. Fifteen of these results are calculated from other numbers in the studies. Numbers for "usually" or "regularly" are combined with numbers from "always" or "almost always" to give the upper half of the range.

10. In this calculation, the "sometimes" category includes: "less than half the time" (Butler), "33% of the time" (Levine and Yost), and "20–49% of sex acts" (Hamilton).

11. Included in this category are instances of "difficulty in reaching orgasm" (Frank et al.), "formerly, not now" (Dickinson and Beam), "1–3 in whole life" (Hamilton), "very rarely" (Schnabl), and "once in a while" (Tavris and Sadd).

### 3. Pair-Bond Accounts of Female Orgasm

1. Donald Symons emphasizes the importance of describing the particular environmental features to which the adaptation is responding (1990, p. 428).

2. George Williams emphasizes this last point, in his discussion about the testing of adaptationist hypotheses (1985, p. 18).

3. Estrus is defined both physiologically, as the time when a certain hormone, luteinizing hormone, or LH, surges and fertility is highest, and behaviorally, as the time when females seek out sexual contacts eagerly. For example, cats and dogs go into "heat," or estrus, and are fertile only during those times, when they are eager to mate. Most of the studies I consider define estrus as being the time that genital swellings appear in nonhuman primates, a rough approximation to the time of peak fertility.

4. And it is no longer a conventional explanation for copulations during estrus, given the theory of infanticide prevention (see Hrdy 1981, pp. 172–177).

5. Linda Wolfe argues that, anatomically, many monkeys differ from human beings in that their clitorises are directly stimulated during intercourse (1991). However, observations of nonhuman primate female orgasm indicate that it occurs primarily outside of intercourse.

6. Note that Hamburg's account is challenged by the existence of chimpanzee orgasm (Lemmon and Allen 1978). Perhaps that was discounted because the orgasm did not occur during intercourse. That might also be the reason that other evidence of nonhuman female orgasm, much of which occurs in female-female encounters, is considered by Hamburg to be "sporadic."

7. The notion of premature ejaculation is not universally well regarded, because it seems to be defined in terms of the timing of female orgasm, but I must refer to it here, as part of the quote from Rancour-Laferriere.

8. Erectile difficulties are estimated at between 3 to 9% of the general population of men, including older men, whereas premature ejaculation is estimated at between 36 to 38% (Spector and Carey 1990, p. 401).

### 4. Further Evolutionary Accounts of Female Orgasm

1. Allen and Lemmon do not claim that their hypothesis is original; they cite C. P. Stone (in Rosvold 1954, p. 96) as having made a similar suggestion (1981, p. 24).

2. Freud tied the capacity to have orgasm with intercourse to women's psychological maturity. Others have since linked it to sexual guilt (Kelly et al. 1990; Davidson and Moore 1994), difficulties with separation (Fisher 1973), self-blame (Loos et al. 1987), conservative attitudes (Derogatis et al. 1986), greater dependency (Terman 1951), and a number of other traits. A recent examination of these claims concludes, "Overall, associations between female orgasm response and psychopathological adjustment have not been supported" (Mah and Binik 2001, p. 834). Support for this conclusion comes from Fisher (1973) and Raboch and Raboch (1992).

3. This trait would seem not to be undermined by the unrepresentative nature of Masters and Johnson's sample, but we cannot be sure.

4. Hrdy's predilection for adaptive accounts of traits has been criticized by Donna Haraway (1989, chap. 15).

### 5. The Byproduct Account

1. Although there is considerable evidence for female ejaculation, the topic is not uncontroversial. Evidence for female ejaculation can be found in Zaviacic et al. (1988), Whipple et al. (1994), and Zaviacic and Whipple (1993). Arguments against female ejaculation are presented in Goldberg et al. (1983) and Alzate (1985b). As Marlene Zuk points out, male orgasm may be inefficient in that multiple intromissions are involved in many species, as well as repeated thrusting. Perhaps male orgasm is not the adaptation that it's cracked up to be (2002, p. 145).

2. As Kinsey said, "Males would be better prepared to understand females, and females to understand males, if they realized that they are alike in their basic anatomy and physiology" (1953, p. 641).

3. Kinsey et al. 1948, pp. 573, 575; Kinsey et al. 1953, p. 587. It is unknown how widespread male nipple sensitivity is.

4. Kinsey et al. 1953, p. 358. They also remark, "At some social levels in our own culture, and in some cultures elsewhere in the world, there may be studied avoidance of pre-coital play, and a social mandate that

sexual contacts be limited to genital unions which were directly carried through to orgasm for the male, with little if any attempt to arouse the female sexually" (1953, p. 361).

5. Masters and Johnson's nonrepresentative sample makes no difference to the validity of these results, since they are based on noncoital as well as coital orgasms.

6. Actually, it is not that the evidence Symons cites is not compelling; he does not cite any evidence at all.

7. Frayser endorses Hrdy's approach to female sexuality. "Females express their sexual assertiveness and awareness of social networks by their participation in nonreproductive (and highly stimulating) sexual behavior" (1985, p. 43).

8. "Some evolutionary psychologists believe that women didn't even evolve their own orgasms; it seems we just got lucky because it was so important for men to seek constant sexual gratification (and in the fetus the clitoris and the glans of the penis develop from the same embryonic phallus)" (1997, p. 404).

9. Donna Haraway delivers an intriguing and sensitive analysis of the relation between orgasm in evolutionary theory and the notion of female agency (1989, pp. 354–367).

### 6. Warring Approaches to Adaptation

1. Kinsey et al. 1948, pp. 573, 575; Kinsey et al. 1953, p. 587 (see the citations there).

2. According to Maynard Smith, "the 'function' of an organ is taken to mean those of its effects which have been responsible for its evolution by natural selection" (1978, p. 23). See the definitions of function in Godfrey-Smith 1994, Burian 1992, and Allen et al. 1998. Note that references to current contributions to fitness are omitted.

3. Burian provides a clear explanation of ways in which the historical definition of adaptation and the definition using current-fitness consequences differ. As Burian puts it, "The fact that possession of a trait complex correlated with increased adaptedness or inclusive fitness of its bearers is not sufficient to establish that it is an (evolutionary) adaptation, for that fact leaves unresolved questions regarding the historical pathway by which the trait arose" (1992, p. 11). Burian and Reeve and Sherman contrast two fundamental approaches to defining adaptation:

one that connects it primarily to "adaptedness"—"its relative fitness or likelihood of reproductive survival" in the current environment—and one that takes a historical approach, emphasizing the shaping of the trait by past evolutionary forces of selection (Burian 1992, p. 8).

4. Phylogenetic inertia involves the persistence of traits through descendant species in the lineage, even in the absence of selection.

5. Note that the existence of the byproduct account does not shut down the sexology research into the possible connection between orgasm and fertility, contrary to what is implied by Alcock and Sherman.

6. Gould and Vrba confuse the issue in one place, in which they describe aptations as consisting of "two partially overlapping subsets: the subset of adaptations and the subset of exaptations" (1982, p. 6). This is an incorrect description of the logic of the terms they set forth; the subsets are not overlapping, as can be seen from their diagram (1982, p. 5).

7. For discussion of and reference to current thinking on this topic, see Larson and Losos 1996 and Reznick and Travis 1996.

## 7. Sperm-Competition Accounts

1. "Capillary forces generated within the mucus and a difference in hydrostatic pressure between the vagina and the peritoneal cavity have been described (Fox et al., 1970); it is, however, unlikely that these factors account for the immediate uptake from the vagina and the directed transport" (Wildt et al. 1998, p. 664). It seems that ongoing peristaltic contractions are responsible for the movement, according to Wildt and colleagues.

2. The follicular phase is the segment of the menstrual cycle after menstruation and up to ovulation. The luteal phase immediately follows ovulation.

3. The proponents of uterine upsuck could argue that increased fertility, while not associated with reproductive success in the past, is currently so associated. This would require that increased fertility today be correlated with increased reproductive success, a connection that has not been demonstrated.

4. These conditions include comparatively large testes; large epididymis; large ejaculate volume; high sperm concentration; high sperm motility; large sperm reserves (capacity for matings per day); large baculum; co-

agulating sperm plugs; post-ejaculatory intromission (with the exception of chimpanzees); and brief copulations (for example, for a chimpanzee, 7 seconds; for a bonobo, 13 seconds) (Richard Wrangham, personal communication).

5. Baker and Bellis hypothesize that the mechanism of uterine upsuck has its effect primarily in sucking sperm into the cervical walls, where they are stored in crypts. Note that this hypothesized mechanism is in tension with the findings from the oxytocin studies, where a semen-like fluid is propelled rapidly through the cervix and toward the back of the uterus.

6. Against Thornhill et al. 1995, Dixson points out that "none of these traits is sexually selected and none is likely to be important for male attractiveness or female mate choice" (1998, p. 200).

7. Numerous other measures were included in their multiple regression analysis, including predicted future earnings, socioeconomic status, self-professed love, and sexual behavior and attitudes.

## 8. Bias

1. William Wimsatt (1987), Nancy Cartwright and S. Nordby (1983), and Cartwright (1983) have argued that false assumptions may be very useful in model building. I agree with them in general, but they are concerned with formal models, and adaptive explanations of a trait are not formal models like the ones we find in population genetics (Lloyd 1983). In adaptive models, if part of the story the researcher tells is false, the explanation itself is considered false. Adaptationist explanations always allow alternative assumptions to be incorporated, however.

2. Once again, if phenotypic plasticity is involved, the situation regarding variability is complicated. Nevertheless, since no one has offered an account appealing to phenotypic plasticity, Symons's interpretation remains plausible.

3. But in a wider sense androcentrism always does harm, because we have no idea in which contexts it leads us astray and in which it doesn't (Longino 1990; Antony 1993; Campbell 2001).

4. Kuhn 1970; Longino 1990, 1993a, 1995b, 2002; Rooney 1992; Nelson and Nelson 1994; Campbell 1998, 2001, among others.

5. "Contextual or social values [some background assumptions] are not

just negative features in inquiry, but can have a positive role in grounding criticism of background assumptions and in fostering the development of empirical investigation in directions it would not otherwise go" (Longino 2002, p. 51).

6. Hempel 1988; Dietrich 1993; Kitcher 1993; Potter 1996; Longino 2002, pp. 63–64, 124–128.

7. Mark Kaplan defends the standard of total evidence against recent naturalist attacks (1994, 2002).

8. Various social approaches to knowledge are discussed in Schmitt (1994). Catherine Elgin (1996) develops a "constructionalist" epistemology in which she discusses the social nature of knowledge, as well as explaining how we can be critical about our own standards, using the notion of reflective equilibrium. She does not specifically address the details of scientific standards and how they can be challenged.

9. "It's likely that something so widely present in animals has some adaptive function" (Rodgers 2001, p. 323).

10. With the exception of assumption 1, which involves adaptationism itself. Still, under the view of all but the most ardent adaptationists, the standard of evidence I use in evaluating assumption 1 is acceptable.

# Bibliography

Addiego, F., E. G. Belzer Jr., J. Comolli, W. Moger, J. D. Perry, and B. Whipple. 1981. Female ejaculation: a case study. *Journal of Sex Research* 17:13–21.

Akers, J. S., and C. H. Conaway. 1979. Female homosexual behavior in *Macaca mulatta*. *Archives of Sexual Behavior* 8 (1): 63–80.

Alcock, J. 1980. Beyond the sociobiology of sexuality: predictive hypotheses. *Behavioral and Brain Sciences* 3:181–182.

———1987. Ardent adaptationism. *Natural History* 96 (4): 4.

———1998. Unpunctuated equilibrium in the *Natural History* essays of Stephen Jay Gould. *Evolution and Human Behavior* 19:321–336.

———2000. Misbehavior—how Stephen Jay Gould is wrong about evolution. *Boston Review* 25 (2): 1–6.

———2001. *The Triumph of Sociobiology*. New York: Oxford University Press.

Alcock, J., and P. Sherman. 1994. The utility of the proximate-ultimate dichotomy in ethology. *Ethology* 96:58–62.

Alexander, R., and K. M. Noonan. 1979. Concealment of ovulation, parental care, and human social evolution. In *Evolutionary Biology and Human Social Behavior: An Anthropological Perspective*, ed. N. Chagnon and W. Irons, pp. 436–453. North Scituate, MA: Duxbury Press.

Alexander, R. D., J. L. Hoogland, R. D. Howard, M. Noonan, and P. W. Sherman. 1979. Sexual dimorphisms and breeding systems in pinnipeds, ungulates, primates, and humans. In *Evolutionary Biology and Human Social Behavior: An Anthropological Perspective*, ed.

N. Chagnon and W. Irons, pp. 402–435. North Scituate, MA: Duxbury Press.

Allen, C., M. Bekoff, and G. V. Lauder, eds. 1998. *Nature's Purposes: Analyses of Function and Design in Biology.* Cambridge: MIT Press.

Allen, J. L., and W. B. Lemmon. 1981. Orgasm in female primates. *American Journal of Primatology* 1:15–34.

Allen, M. L. 1977. Sexual response and orgasm in the female chimpanzee *(Pan troglodytes).* Master's thesis: University of Oklahoma.

Allison, A. C. 1954. The distribution of the sickle-cell trait in East Africa and elsewhere, and its apparent relationship to the incidence of subtertian malaria. *Transactions of the Royal Society of Tropical Medicine and Hygiene* 48:312–318.

——1957. Parasitological reviews: malaria in carriers of the sickle-cell trait and in newborn children. *Experimental Parasitology* 8:418–447.

Altshuler, S. 1986. The hypothesis of female ejaculation: too little interest, too little research. *Journal of Social Work and Human Sexuality* 4 (1/2): 125–139.

Alzate, H. 1985a. Vaginal eroticism and female orgasm: a current appraisal. *Journal of Sex and Marital Therapy* 11:271–284.

——1985b. Vaginal eroticism: a replication study. *Archives of Sexual Behavior* 14: 529–537.

Amberson, J. L., and P. W. Hoon. 1985. Hemodynamics of sequential orgasm. *Archives of Sexual Behavior* 14:351–360.

Anastasi, A. 1988. *Psychological Testing.* New York: Macmillan.

Anderson, E. 1995. Knowledge, human interests, and objectivity in feminist epistemology. *Philosophical Topics* 23 (2): 27–58.

Andrews, P. W., S. W. Gangestad, and D. Matthews. 2002. Adaptationism—how to carry out an exaptationist program. *Behavioral and Brain Sciences* 25 (4): 489–504.

Antony, L. M. 1993. Quine as feminist: the radical import of naturalized epistemology. In *A Mind of One's Own: Feminist Essays on Reason and Objectivity,* ed. L. M. Antony and C. Witt, pp. 185–226. Boulder: Westview Press.

Armstrong, D. P. 1991. Levels of cause and effect as organizing principles for research in animal behaviour. *Canadian Journal of Zoology* 69:823–829.

Arnold, S. J. 1983. Morphology, performance, and fitness. *American Zoologist* 23:347–361.

Baker, R. R., and M. A. Bellis. 1993a. Human sperm competition: ejaculate adjustment by males and the function of masturbation. *Animal Behaviour* 46:861–885.

———1993b. Human sperm competition: ejaculate manipulation by females and a function for the female orgasm. *Animal Behaviour* 46:887–909.

———1995. *Human Sperm Competition*. London: Chapman and Hall.

Bancroft, J. 1989. *Human Sexuality and Its Problems*. New York: Churchill Livingstone.

Barash, D. 1977. *Sociobiology and Behavior*. New York: Elsevier North-Holland.

Barrett, L., R. Dunbar, and J. Lycett. 2002. *Human Evolutionary Psychology*. Princeton: Princeton University Press.

Bartlett, R. G. 1956. Physiologic responses during coitus. *Journal of Applied Physiology* 9:469–472.

Beach, F. A. 1968. Factors involved in the control of mounting behavior by female mammals. In *Reproduction and Sexual Behavior*, ed. M. Diamond, pp. 83–131. Bloomington: Indiana University Press.

———1973. Human sexuality and evolution. In *Advances in Behavioral Biology*, ed. W. Montagna and W. A. Sadler, pp. 333–365. New York: Plenum Press.

———1976. Sexual attractivity, proceptivity and receptivity in female primates. *Hormones and Behavior* 1 (7): 105–138.

Beauchamp, G. K., K. Yamazaki, J. Baird, and E. A. Boyse. 1988. Preweaning experience in the control of mating preferences by genes in the major histocompatibility complex of the mouse. *Behavior Genetics* 18:537–547.

Beer, A. E., A. E. Semprini, X. Y. Zhu, and J. F. Quebbeman. 1985. Pregnancy outcome in human couples with recurrent spontaneous abortions: HLA antigen profiles, HLA antigen sharing, female serum MLR blocking factors, and paternal leukocyte immunization. *Experimental Clinical Immunogenetics* 2:137–153.

Bernds, W. P., and D. Barash. 1979. Early termination of parental investment in mammals, including humans. In *Evolutionary Biology and Human Social Behavior: An Anthropological Perspective*, ed. N.

Chagnon and W. Irons, pp. 487–505. North Scituate, MA: Duxbury Press.

Birkhead, T. R. 1998. Cryptic female choice: criteria for establishing female sperm choice. *Evolution* 52:1212–1218.

———2000. *Promiscuity: An Evolutionary History of Sperm Competition*. Cambridge: Harvard University Press.

Birkhead, T. R., and F. M. Hunter. 1990. Mechanisms of sperm competition. *Trends in Ecology and Evolution* 5 (2): 48–52.

Blaicher, W., D. Gruber, C. Bieglmayer, A. M. Blaicher, W. Knogler, and J. C. Huber. 1999. The role of oxytocin in relation to female sexual arousal. *Gynecologic and Obstetric Investigation* 47:125–126.

Bohlen, J. G., J. P. Held, M. O. Sanderson, and C. M. Boyer. 1982. Development of a woman's multiple orgasm pattern: a research case report. *Journal of Sex Research* 18:130–145.

Bolis, P. F., V. Soro, M. Martinetti Bianchi, and M. Belvedere. 1985. HLA compatibility and human reproduction. *Clinical Experimental Obstetrics and Gynecology* 12:9–12.

Boyse, E. A., G. K. Beauchamp, J. Bard, and K. Yamazaki. 1990. Behavior and the major histocompatibility complex of the mouse. In *Psychoneuroimmunology*, ed. R. Ader, D. L. Felten, and N. Cohen, pp. 831–845. London: Academic Press.

Brandon, R. N. 1990. *Adaptation and Environment*. Princeton: Princeton University Press.

Burian, R. M. 1983. Adaptation. In *Dimensions of Darwinism*, ed. M. Grene, pp. 287–314. Cambridge: Cambridge University Press.

———1992. Adaptation. In *Keywords in Evolutionary Biology*, ed. E. F. Keller and E. A. Lloyd, pp. 7–12. Cambridge: Harvard University Press.

Burton, F. D. 1971. Sexual climax in female *Macaca mulatta*. In *Proceedings, Third International Congress of Primatology*, ed. H. Kummer, vol. 3, pp. 180–191. Basel: S. Karger.

Buss, D. M. 1994. *The Evolution of Desire: Strategies of Human Mating*. New York: Basic Books.

Butler, C. A. 1976. New data about female sexual response. *Journal of Sex and Marital Therapy* 2 (1): 40–46.

Campbell, B. 1966. *Human Evolution: An Introduction to Man's Adaptations*. Chicago: Aldine.

Campbell, R. 1998. *Illusions of Paradox: A Feminist Epistemology Naturalized.* New York: Rowman and Littlefield.

———2001. The bias paradox in feminist epistemology. In *Engendering Rationalities,* ed. S. Morgan and N. Tuana, pp. 195–217. Stony Brook, NY: SUNY Press.

Carnap, R. 1962. *Logical Foundations of Probability.* Chicago: University of Chicago Press.

Cartwright, N. 1983. *How the Laws of Physics Lie.* Oxford: Oxford University Press.

Cartwright, N., and S. Nordby. 1983. How approximations take us away from theory and toward truth. *Pacific Philosophical Quarterly* 64:273–280.

Caulfield, M. D. 1985. Sexuality in human evolution: what is natural in sex? *Feminist Studies* 11 (2): 343–363.

Cavalli-Sforza, L. L., and W. F. Bodmer. 1971. *The Genetics of Human Populations.* San Francisco: W. H. Freeman.

Chagnon, N., and W. Irons, eds. 1979. *Evolutionary Biology and Human Social Behavior: An Anthropological Perspective.* North Scituate, MA: Duxbury Press.

Chan, D. W. 1990. Sex knowledge, attitudes, and experience of Chinese medical students in Hong Kong. *Archives of Sexual Behavior* 19 (1): 73–93.

Chesser, E. 1956. *The Sexual, Marital, and Family Relationships of the English Woman.* London: Hutchinson's Medical Publications.

Chevalier-Skolnikoff, S. 1974. Male-female, female-female, and male-male sexual behavior in the stumptail monkey. *Archives of Sexual Behavior* 3 (2): 95–116.

———1976. Homosexual behavior in a laboratory group of stumptail monkeys: forms, contexts, and possible social functions. *Archives of Sexual Behavior* 5 (6): 511–527.

Christensen, E., and P. Hertoft. 1980. Sexual activity and attitude during pregnancy and the postpartum period. In *Medical Sexology,* ed. R. Forleo and W. Pasini, pp. 357–364. Littleton, MA: PSG Publishing Co.

Clement, U., G. Schmidt, and M. Kruse. 1984. Changes in sex differences in sexual behavior: a replication of a study on West German students, 1966–1981. *Archives of Sexual Behavior* 13 (2): 99–120.

Clifford, R. E. 1978. Subjective sexual experience in college women. *Archives of Sexual Behavior* 7 (3): 183–197.

Clutton-Brock, T. H., and P. H. Harvey. 1979. Comparison and adaptation. *Proceedings of the Royal Society of London* B 205:547–565.

Cochran, W. G., F. Mosteller, and J. W. Tukey. 1953. Statistical problems of the Kinsey report. *Journal of the American Statistical Association* 48 (264): 673–716.

Crocker, W., and J. Crocker. 1994. *The Canela: Bonding through Kinship, Ritual, and Sex.* Fort Worth: Harcourt Brace College Publishers.

Crook, J. H. 1972. Sexual selection, dimorphism, and social organization in the primates. In *Sexual Selection and the Descent of Man, 1871–1971,* ed. B. Campbell, pp. 231–281. Chicago: Aldine.

Daly, M., and M. I. Wilson. 1999. Review: human evolutionary psychology and animal behaviour. *Animal Behaviour* 57:509–519.

Darling, C. A., J. K. Davidson, and C. Conway-Welch. 1990. Female ejaculation: perceived origins, the Grafenberg spot/area, and sexual responsiveness. *Archives of Sexual Behavior* 19 (1): 29–47.

Darling, C. A., J. K. Davidson, and D. A. Jennings. 1991. The female sexual response revisited: understanding the multiorgasmic experience in women. *Archives of Sexual Behavior* 20:527–540.

Darwin, C. 1964. *On the Origin of Species.* Cambridge: Harvard University Press.

Davenport, W. H. 1977. Sex in cross-cultural perspective. In *Human Sexuality in Four Perspectives,* ed. F. Beach, pp. 115–163. Baltimore: Johns Hopkins University Press.

Davidson, J. K., and N. B. Moore. 1994. Guilt and lack of orgasm during sexual intercourse: myth versus reality among college women. *Journal of Sex Education and Therapy* 20:153–174.

Derogatis, L. R., P. J. Fagan, C. W. Schmidt, T. N. Wise, and K. S. Gildeu. 1986. Psychological subtypes of anorgasmia: a marker variable approach. *Journal of Sex and Marital Therapy* 12:197–210.

de Waal, F. B. M. 1995. Sex as an alternative to aggression in the bonobo. In *Sexual Nature, Sexual Culture,* ed. P. R. Abramson and S. D. Pinkerton, pp. 37–56. Chicago: University of Chicago Press.

Dewsbury, D. A. 1980. Methods in the two sociobiologies. *Behavioral and Brain Sciences* 3:183–184.

———1992. On the problems studied in ethology, comparative psychology, and animal behavior. *Ethology* 92:89–107.

Diamond, M. 1980. The biosocial evolution of human sexuality, a comment on Donald Symons' precis of the evolution of human sexuality. *Behavioral and Brain Sciences* 3:184–186.

Dickinson, R. L., and L. Beam. 1931. *A Thousand Marriages: A Medical Study of Sex Adjustment.* Baltimore: Williams and Wilkins.

Dietrich, M. 1993. Underdetermination and the limits of interpretative flexibility. *Perspectives on Science* 1:109–126.

Dixson, A. F. 1998. *Primate Sexuality: Comparative Studies of the Prosimians, Monkeys, Apes, and Human Beings.* Oxford: Oxford University Press.

Domjan, M. 1998. *The Principles of Learning and Behavior.* Pacific Grove, CA: Brooks/Cole Publishing.

Duty, S. M., M. J. Silva, D. B. Barr, J. W. Brock, L. Ryan, Z. Chen, R. F. Herrick, D. C. Christiani, and R. Hauser. 2003. Phthalate exposure and human semen parameters. *Epidemiology* 14:269–277.

Eberhard, W. G., and C. Cordero. 1995. Sexual selection by cryptic female choice on male seminal products: a new bridge between sexual selection and reproductive physiology. *Trends in Ecology and Evolution* 10:493–495.

Egid, K., and J. L. Brown. 1989. The major histocompatibility complex and female mating preferences in mice. *Animal Behaviour* 38:548–549.

Eibl-Eibesfeldt, I. 1970. *Ethology: The Biology of Behavior.* New York: Holt, Rinehart, and Winston.

Eklund, A. 1997. The major histocompatibility complex and mating preferences in wild house mice *(Mus domesticus)*. *Behavioral Ecology* 8:630–634.

Eklund, A., K. Egid, and J. L. Brown. 1991. The major histocompatibility complex and mating preferences of male mice. *Animal Behaviour* 42:693–694.

Elgin, C. Z. 1996. *Considered Judgement.* Princeton: Princeton University Press.

Endler, J. A. 1986. *Natural Selection in the Wild.* Princeton: Princeton University Press.

Falconer, D. S. 1981. *Introduction to Quantitative Genetics.* New York: Longman.

Faulkner, W. 1980. The obsessive orgasm: science, sex, and female sexuality. In *Alice through the Microscope: The Power of Science over*

*Women's Lives,* by the Brighton Women and Science Group, pp. 139–162. London: Virago.

Fausto-Sterling, A., P. Gowaty, and M. Zuk. 1997. Evolutionary psychology and Darwinian feminism. *Feminist Studies* 23 (2): 403–417.

Ferguson, G. W., and S. F. Fox. 1984. Annual variation of survival advantage of large juvenile side-blotched lizards, *Uta stansburiana:* its causes and evolutionary significance. *Evolution* 38:342–349.

Fisher, S. 1973. *The Female Orgasm: Psychology, Physiology, Fantasy.* New York: Basic Books.

Ford, C. S., and F. Beach. 1951. *Patterns of Sexual Behavior.* New York: Harper.

Fox, C. A., and B. Fox. 1971. A comparative study of coital physiology with special reference to the sexual climax. *Journal of Reproduction and Fertility* 24:319–336.

Fox, C. A., H. S. Wolff, and J. A. Baker. 1970. Measurement of intra-vaginal and intra-uterine pressures during human coitus by radio-telemetry. *Journal of Reproduction and Fertility* 22:243–251.

Fox, R. 1993. Male masturbation and female orgasm. *Society* 30 (6): 21–25.

Frank, E., C. Anderson, and D. Rubinstein. 1978. Frequency of sexual dysfunction in normal couples. *New England Journal of Medicine* 299 (3): 111–115.

Frayser, S. 1985. *Varieties of Sexual Experience: An Anthropological Perspective on Human Sexuality.* New Haven: HRAF Press.

Freeman, S., and J. C. Herron. 1998. *Evolutionary Analysis.* Upper Saddle River, NJ: Prentice Hall.

Freud, S. 1905. *Three Essays on the Theory of Sexuality.* New York: Basic Books.

Gallup, G. G., and S. D. Suarez. 1983. Optimal reproductive strategies for bipedalism. *Journal of Human Evolution* 12:193–196.

Gebhard, P. H. 1966. Factors in marital orgasm. *Journal of Social Issues* 22:88–95.

———1970. The sexuality of women. In *The Sexuality of Women,* ed. P. H. Gebhard, J. Raboch, and H. Giese, pp. 1–44. New York: Stein and Day.

Gebhard, P. H., J. Raboch, and H. Giese, eds. 1970. *The Sexuality of Women.* The Library of Sexual Behavior. New York: Stein and Day.

Geertz, C. 1990. A lab of one's own. *New York Review of Books* 37 (November 8): 19.

Godfrey-Smith, P. 1994. A modern history theory of functions. *Nous* 28: 344–362.

———2001. Three kinds of adaptationism. In *Adaptationism and Optimality,* ed. S. H. Orzack and E. Sober, pp. 335–357. Cambridge: Cambridge University Press.

Goldberg, D. C., B. Whipple, R. E. Fishkin, H. Waxman, P. J. Fink, and M. Weisberg. 1983. The Grafenberg spot and female ejaculation: a review of initial hypotheses. *Journal of Sex and Marital Therapy* 9:27–37.

Goldfoot, D. A., A. K. Slob, G. Scheffler, J. A. Robinson, S. J. Wiegand, and J. Cords. 1975. Multiple ejaculations during prolonged sexual tests and lack of resultant serum testosterone increases in male stumptail macaques *(M. arctoides). Archives of Sexual Behavior* 4 (5): 547–560.

Goldfoot, D. A., H. Westerborg-van Loon, W. Groeneveld, and A. Koos Slob. 1980. Behavioral and physiological evidence of sexual climax in the female stump-tailed macaque *(Macaca arctoides). Science* 208:1477–1478.

Goldman, A. 1995. Psychological, social, and epistemic factors in the theory of science. In *Proceedings of the Biennial Meeting of the Philosophy of Science Association, 1994,* ed. R. Burian, M. Forbes, and D. Hull, pp. 277–286. East Lansing, MI: Philosophy of Science Association.

Goodall, J. 1986. *The Chimpanzees of Gombe: Patterns of Behavior.* Cambridge: Harvard University Press.

Gould, S. J. 1987a. Freudian slip. *Natural History* 96 (2): 14–21.

———1987b. Reply to Alcock. *Natural History* 96 (4): 4–6.

Gould, S. J., and E. Vrba. 1982. Exaptation—a missing term in the science of form. *Paleobiology* 8:4–15.

Goy, R. W., and D. A. Goldfoot. 1975. Neuroendocrinology: animal models and problems of human sexuality. *Archives of Sexual Behavior* 4 (4): 405–420.

Grafen, A. 1988. On the uses of data on lifetime reproduction. In *Reproductive Success: Studies on Individual Variation in Contrasting Breeding Systems,* ed. T. H. Clutton-Brock, pp. 454–471. Chicago: University of Chicago Press.

278 · Bibliography

Grafenberg, E. 1950. The role of urethra in female orgasm. *The International Journal of Sexology* 3:145–148.

Grant, P. R. 1986. *Ecology and Evolution of Darwin's Finches*. Princeton: Princeton University Press.

Griffiths, A. J. F., W. M. Gelbart, R. Lewontin, and J. H. Miller. 2002. *Modern Genetic Analysis: Integrating Genes and Genomes*. New York: W. H. Freeman.

Gross, P. R., and N. Levitt. 1994. *Higher Superstition: The Academic Left and Its Quarrels with Science*. Baltimore: Johns Hopkins University Press.

Haack, S. 1993. *Evidence and Inquiry: Towards Reconstruction in Epistemology*. Oxford: Blackwell.

Hafez, E. S. E. 1971. Reproductive cycles. In *Comparative Reproduction of Nonhuman Primates,* ed. E. S. E. Hafez, pp. 160–204. Springfield, IL: Charles C. Thomas.

Hamburg, B. A. 1978a. The biosocial bases of sex difference. In *Human Evolution: Biosocial Perspectives,* ed. S. L. Washburn and E. R. McCown, pp. 154–213. Menlo Park, CA: Benjamin/Cummings Publishing Co.

———1978b. The psychobiology of sex differences: an evolutionary perspective. In *Sex Differences in Behavior,* ed. R. C. Friedman, pp. 373–392. Huntington, NY: R. E. Krieger.

Hamilton, G. V. 1929. *A Research in Marriage*. New York: Albert & Charles Boni.

Hanby, J. 1976. Sociosexual development in primates. In *Perspectives in Ethology,* ed. P. P. G. Bateson, pp. 1–67. New York: Plenum Press.

Hanby, J., L. Robertson, and C. Phoenix. 1971. Sexual behavior in a confined troop of Japanese macaques. *Folia Primatologica* 16:123–143.

Haraway, D. 1989. *Primate Visions: Gender, Race, and Nature in the World of Modern Science*. London: Routledge.

———1991. *Simians, Cyborgs, and Women: The Reinvention of Nature*. London: Routledge.

Hartman, W. E., and M. A. Fithian. 1994. Physiological response patterns of 751 reseach volunteers. Paper presented at the Society for the Scientific Study of Sexuality Western Region Conference, San Diego, April 1994.

Heiman, J. R. 1980. Selecting for a sociobiological fit. *Behavioral and Brain Sciences* 3:189–190.

Hempel, C. G. 1988. Provisos: a problem concerning the inferential functions of scientific laws. In *The Limits of Deductivism,* ed. A. Grünbaum and W. Salmon, pp. 19–36. Berkeley: University of California Press.

Heyn, A. 1921. Studien zur Physiologie des Geschlechtslebens der Frau. *Geschlecht und Gesellschaft* 10:405–408.

Hite, S. 1976. *The Hite Report: A Nationwide Survey of Female Sexuality.* New York: MacMillan.

———1981. *The Hite Report on Male Sexuality.* New York: Ballantine Books.

Ho, H. N., T. J. Gill III, R. P. Nsieh, H. J. Hsieh, and T. Y. Lee. 1990. Sharing of human leukocyte antigens in primary and secondary recurrent spontaneous abortions. *American Journal of Obstetrics and Gynecology* 163:178–188.

Hoch, Z. 1980. The sensory arm of the female orgasmic [*sic*] reflex. *Journal of Sex Education and Therapy* 6:4–7.

Howell, N. 1979. *Demography of the Dobe !Kung.* New York: Academic Press.

Hrdy, S. B. 1979. The evolution of human sexuality: the latest word and the last. *Quarterly Review of Biology* 54 (September): 309–314.

———1981. *The Woman That Never Evolved.* Cambridge: Harvard University Press.

———1988. The primate origins of human sexuality. In *The Evolution of Sex,* ed. R. Bellig and G. Stevens, pp. 101–136. San Francisco: Harper & Row.

———1996. The evolution of female orgasms: logic please but no atavism. *Animal Behaviour* 52: 851–852.

———1997. Raising Darwin's consciousness: female sexuality and the prehominid origins of patriarchy. *Human Nature* 8 (1): 1–49.

———1999. *Mother Nature: Maternal Instincts and How They Shape the Human Species.* New York: Ballantine Books.

Huey, C. J., G. Kline-Graber, and B. Graber. 1981. Time factors and orgasmic response. *Archives of Sexual Behavior* 10 (2): 111–118.

Hunt, M. 1974. *Sexual Behavior in the 1970's.* Chicago: Playboy Press.

Hurlbert, D. F., and C. Apt. 1995. The coital alignment technique and directed masturbation: a comparative study on female orgasm. *Journal of Sex and Marital Therapy* 21 (1): 21–29.

Irons, W. 1983. Human female reproductive strategies. In *The Social Be-*

*havior of Female Vertebrates,* ed. S. K. Wasser, pp. 169–213. New York: Academic Press.

Jacob, S., M. K. McClintock, B. Zelano, and C. Ober. 2002. Paternally inherited HLA alleles are associated with women's choice of male odor. *Nature Genetics* 30:175–179.

Jamieson, I. G. 1989. Levels of analysis or analyses at the same level. *Animal Behaviour* 37 (4): 696–697.

Javert, C. T. 1957. *Spontaneous and Habitual Abortion.* New York: McGraw-Hill.

Jayne, B. C., and A. F. Bennett. 1990. Selection on locomotor performance capacity in natural populations of garter snakes. *Evolution* 44:1204–1229.

Jennions, M. D. 1997. Female promiscuity and genetic incompatibility. *Trends in Ecology and Evolution* 12:251–252.

Jolly, A. 2001. *Lucy's Legacy: Sex and Intelligence in Human Evolution.* Cambridge: Harvard University Press.

Kano, T. 1980. Social behavior of wild pygmy chimpanzees of Wamba: a preliminary report. *Journal of Human Evolution* 9:243–260.

———1982. The social group of pygmy chimpanzees *(Pan paniscus)* of Wamba. *Primates* 23 (2): 171–188.

———1992. *The Last Ape: Pygmy Chimpanzee Behavior and Ecology.* Stanford: Stanford University Press.

Kaplan, H. S. 1974. *The New Sex Therapy: Active Treatment of Sexual Dysfunctions.* New York: Brunner/Mazel.

Kaplan, M. 1994. Epistemology denatured. In *Midwest Studies in Philosophy XIX: Studies in Philosophical Naturalism,* ed. P. A. French, T. E. Uehling, and K. Wettstein, pp. 350–365. Notre Dame, IN: University of Notre Dame Press.

———2002. Decision theory and epistemology. In *The Oxford Handbook of Epistemology,* ed. P. K. Moser, pp. 434–462. Oxford: Oxford University Press.

Karl, A., G. Metzner, H. J. Seewald, M. Karl, U. Born, and G. Tilch. 1989. HLA compatibility and susceptibility to habitual abortion. Results of histocompatibility testing of couples with frequent miscarriages. *Allergie und Immunologie* 35:133–140.

Keller, E. F., and H. E. Longino. 1996. *Feminism and Science.* New York: Oxford University Press.

Kelly, M. P., D. S. Strassberg, and J. R. Kircher. 1990. Attitudinal and experiential correlates of anorgasmia. *Archives of Sexual Behavior* 19:165–177.

Kinsey, A. C., W. Pomeroy, and C. Martin. 1948. *Sexual Behavior in the Human Male*. Philadelphia: W. B. Saunders.

Kinsey, A. C., W. Pomeroy, C. Martin, and P. H. Gebhard. 1953. *Sexual Behavior in the Human Female*. Philadelphia: W. B. Saunders.

Kitcher, P. 1985. *Vaulting Ambition: Sociobiology and the Quest for Human Nature*. Cambridge: MIT Press.

———1993. *The Advancement of Science: Science without Legend, Objectivity without Illusions*. Oxford: Oxford University Press.

Kline-Graber, G., and B. Graber. 1975. *Women's Orgasm: A Guide to Sexual Satisfaction*. Indianapolis: Bobbs-Merrill.

Komisaruk, B. R., and B. Whipple. 1995. The suppression of pain by genital stimulation in females. *Annual Review of Sex Research* 6:151–186.

Kopp, M. E. 1934. *Birth Control in Practice: Analysis of Ten Thousand Case Histories of the Birth Control Clinical Research Bureau*. New York: Robert M. McBride.

Koyama, M., F. Saji, S. Takahashi, M. Takemura, Y. Samejima, T. Kameda, T. Kimura, and O. Tanizawa. 1991. Probabilistic assessment of the HLA sharing of recurrent spontaneous abortion couples in the Japanese population. *Tissue Antigens* 37:211–217.

Kuhn, T. 1970. *The Structure of Scientific Revolutions*. 2nd ed. Chicago: University of Chicago Press.

Ladas, A. K., B. Whipple, and J. D. Perry. 1982. *The G Spot and Other Recent Discoveries about Human Sexuality*. New York: Holt, Rinehart, and Winston.

Laitinen, T. 1993. A set of MHC haplotypes found among Finnish couples suffering from recurrent spontaneous abortions. *American Journal of Reproductive Immunology* 29:148–154.

Lancaster, J. 1975. *Primate Behavior and the Emergence of Human Culture*. New York: Holt, Rinehart, and Winston.

Lancaster, J. B., and C. S. Lancaster. 1980. The division of labor and the evolution of human sexuality. *Behavioral and Brain Sciences* 3:193.

Lande, R., and S. J. Arnold. 1983. The measurement of selection on correlated characters. *Evolution* 37:1210–1226.

Landis, C., A. Landis, M. M. Bolles, H. F. Metzger, M. W. Pitts, D. A. D'Esopo, H. C. Moloy, S. J. Kleegman, and R. L. Dickinson. 1940. *Sex in Development.* New York: Paul B. Hoeber.

Landry, C., D. Garant, P. Duchesne, and L. Bernatchez. 2001. Good genes as heterozygosity: MHC and mate choice in Atlantic salmon *(Salmo salar). Proceedings of the Royal Society of London* B 268:1279–1285.

Larson, A., and J. B. Losos. 1996. Phylogenetic systematics of adaptation. In *Adaptation,* ed. M. R. Rose and G. V. Lauder, pp. 187–221. San Diego: Academic Press.

Laudan, L. 1984. *Science and Values: The Aims of Science and Their Role in Scientific Debate.* Berkeley: University of California Press.

Lauder, G. V. 1996. The argument from design. In *Adaptation,* ed. M. R. Rose and G. V. Lauder, pp. 55–93. San Diego: Academic Press.

Laumann, E. O., J. H. Gagnon, R. T. Michael, and S. Michaels. 1994. *The Social Organization of Sexuality: Sexual Practices in the United States.* Chicago: University of Chicago Press.

Lemmon, W. B., and M. L. Allen. 1978. Continual sexual receptivity in the female chimpanzee *(Pan troglodytes). Folia Primatologica* 30:80–88.

Levin, R. J. 1981. The female orgasm: a current appraisal. *Journal of Psychosomatic Research* 25:119–133.

Levine, S. B., and M. A. Yost, Jr. 1976. Frequency of sexual dysfunction in a general gynecological clinic: an epidemiological approach. *Archives of Sexual Behavior* 5 (3): 229–238.

Levins, R., and R. Lewontin. 1985. *The Dialectical Biologist.* Cambridge: Harvard University Press.

Lewontin, R. 2001. *It Ain't Necessarily So: The Dream of the Human Genome and Other Illusions.* 2nd ed. New York: New York Review of Books.

Lightfoot-Klein, H. 1989. The sexual experience and marital adjustment of genitally circumcised and infibulated females in the Sudan. *Journal of Sex Research* 26 (3): 375–392.

Lindburg, D. G. 1983. Mating behavior and estrus in the Indian rhesus monkey. In *Perspectives in Primate Biology,* ed. P. K. Seth, pp. 45–61. New Delhi: Today and Tomorrow Press.

Lloyd, E. A. 1983. The nature of Darwin's support for the theory of natural selection. *Philosophy of Science* 50:112–129.

————1988/1994. *The Structure and Confirmation of Evolutionary Theory.* Westport, CT: Greenwood Press, 1988; Princeton: Princeton University Press, 1994.

————1993. Pre-theoretical assumptions in evolutionary explanations of female sexuality. *Philosophical Studies* 69:139–153.

————1997. Feyerabend, Mill, and Pluralism. *Philosophy of Science* 64 (4): S396–S408.

————2001. Units and levels of selection: an anatomy of the units of selection debates. In *Thinking about Evolution: Historical, Philosophical, and Political Perspectives,* ed. R. Singh, C. Krimbas, D. Paul, and J. Beatty, pp. 267–291. Cambridge: Cambridge University Press.

Lloyd, E. A., and S. J. Gould. 1993. Species selection on variability. *Proceeedings of the National Academies of Science USA* 90:595–599.

Longino, H. E. 1990. *Science as Social Knowledge: Values and Objectivity in Scientific Inquiry.* Princeton: Princeton University Press.

————1993a. Essential tensions—phase two: feminist, philosophical, and social studies of science. In *A Mind of One's Own: Feminist Essays on Reason and Objectivity,* ed. L. M. Antony and C. Witt, pp. 257–272. Boulder: Westview Press.

————1993b. Subjects, power, and knowledge: description and prescription in feminist philosophies of science. In *Feminist Epistemologies,* ed. L. Alcoff and E. Potter, pp. 101–120. London: Routledge.

————1995. Gender, politics, and the theoretical virtues. *Synthese* 104 (3): 383–397.

————2002. *The Fate of* Knowledge. Princeton: Princeton University Press.

Loos, V. E., C. F. Bridges, and J. W. Critelli. 1987. Weiner's attribution theory and female orgasmic consistency. *Journal of Sex Research* 23 (3): 348–361.

Loy, J. 1970. Peri-menstrual sexual behavior among rhesus monkeys. *Folia Primatologica* 13:286–297.

Mackey, J. P., and F. Vivarelli. 1952. Tanganyika Annual Report of the Medical Laboratory, 1952. Cited in J. P. Mackey and F. Vivarelli. 1954. Sickle-cell anaemia. *British Medical Journal* 1:775–779.

Mah, K., and Y. M. Binik. 2001. The nature of human orgasm: a critical review of major trends. *Clinical Psychology Review* 21 (6): 823–856.

Marshall, D. S. 1971. Sexual behavior on Mangaia. In *Human Sexual Be-*

*havior: Variations in the Ethnographic Spectrum,* ed. D. S. Marshall and R. C. Suggs, pp. 103–162. Englewood Cliffs, NJ: Prentice-Hall.

Marshall, D. S., and R. C. Suggs, eds. 1971. *Human Sexual Behavior: Variations in the Ethnographic Spectrum.* Englewood Cliffs, NJ: Prentice-Hall.

Masters, W. H., and V. E. Johnson. 1965. The sexual response of the human female. In *Sex Research: New Developments,* ed. J. Money, pp. 53–112. New York: Holt, Rinehart, and Winston.

——1966. *Human Sexual Response.* Boston: Little, Brown.

——1970. *Human Sexual Inadequacy.* Boston: Little, Brown.

Mayr, E. 1983. How to carry out the adaptationist program? *American Naturalist* 121:324–334.

Mead, M. 1950. *Male and Female.* Harmondsworth, Eng.: Penguin Books.

Meddis, R. 1984. *Statistics Using Ranks: A Unified Approach.* Oxford: Blackwell.

Michael, R., G. S. Saayman, and D. Zumpe. 1967. Sexual attractiveness and receptivity in rhesus monkeys. *Nature* 215: 554–556.

Michael, R. P., M. I. Wilson, and D. Zumpe. 1974. The bisexual behavior of female rhesus monkeys. In *Sex Differences in Behavior,* ed. R. Friedman, pp. 399–412. New York: Wiley.

Miller, G. S. 1928. Some elements of sexual behavior in primates and their possible influence on the beginings of human social development. *Journal of Mammalogy* 9 (4): 273–293.

Mitchell, S. D. 1992. On pluralism and competition in evolutionary explanations. *American Zoologist* 32:135–144.

Mitchell-Olds, T., and R. G. Shaw. 1987. Regression analysis of natural selection: statistical and biological interpretation. *Evolution* 41:1149–1161.

Moeller, W. 1975. Edentates. In *Grzimek's Animal Life Encyclopedia,* ed. B. Grzimek, vol. 11, pp. 149–181. New York: Van Nostrand Reinhold.

Montgomerie, R. D., and H. Bullock. 1999. Fluctuating asymmetry and the human female orgasm. Paper presented at the 11th annual meeting of the Human Behavior and Evolution Society, Salt Lake City, June 1999.

Mori, A. 1984. An ethological study of pygmy chimpanzees in Wamba, Zaire: a comparison with chimpanzees. *Primates* 25 (3): 255–278.

Morris, D. 1967. *The Naked Ape: A Zoologist's Study of the Human Animal.* New York: McGraw-Hill.

Motulsky, A. G. 1964. Hereditary red cell traits and malaria. *American Journal of Tropical Medicine and Hygiene* 13:147–158.

Nadler, R. D. 1980. Determination of sexuality in the great apes. In *Medical Sexology,* ed. R. Forleo and W. Pasini, pp. 215–222. Littleton, MA: PSG Publishing.

Naples, V. L. 1999. Morphology, evolution, and function of feeding in the giant anteater *(Myrmecophaga tridactyla). Journal of Zoological Research* 249:19–41.

Nelson, J., and L. H. Nelson. 1994. No rush to judgement. *The Monist* 77 (4): 486–509.

Newton, N. 1973. Interrelationships between sexual responsiveness, birth, and breast feeding. In *Contemporary Sexual Behavior: Critical Issues in the 1970s,* ed. J. Zubin and J. Money, pp. 77–98. Baltimore: Johns Hopkins University Press.

Novacek, M. J. 1996. Paleontological data and the study of adaptation. In *Adaptation,* ed. K. D. Rose and G. V. Lauder, pp. 311–363. San Diego: Academic Press.

Ober, C. 1992. The maternal-fetal relationship in human pregnancy: an immunological perspective. *Experimental and Clinical Immunogenetics* 9:1–14.

Ober, C., L. R. Weitkamp, N. Cox, H. Dytch, D. Kostyu, and S. Elias. 1997. HLA and mate choice in humans. *American Journal of Human Genetics* 61: 497–504.

Overstreet, J. W. 1983. Transport of gametes in the reproductive tract of the female mammal. In *Mechanism and Control of Animal Fertilization,* ed. J. F. Hartmann, pp. 499–543. New York: Academic Press.

Parker, G. A. 1970a. Sperm competition and its evolutionary consequences in the insects. *Biological Reviews* 45:525–567.

———1970b. Sperm competition and its evolutionary effect on copula duration in the fly *Scatophaga stercoraria. Journal of Insect Physiology* 16:1301–1328.

Pauling, L., H. A. Itano, S. J. Singer, and I. C. Wells. 1949. Sickle cell anemia, a molecular disease. *Science* 110: 543–548.

Penn, D. J., and W. K. Potts. 1999. The evolution of mating preferences and major histocompatibility complex genes. *The American Naturalist* 153 (2): 145–164.

Perry, J. D., and B. Whipple. 1982. Multiple components of the female orgasm. In *The Circumvaginal Musculature and Sexual Function,* ed. B. Graber, pp. 101–114. New York: Karger.

Pigliucci, M. 2001. *Phenotypic Plasticity: Beyond Nature and Nurture.* Baltimore: Johns Hopkins University Press.

Potter, E. 1995. Good science and good philosophy of science. *Synthese* 104 (5): 423–439.

———1996. Underdetermination undeterred. In *Feminism, Science, and the Philosophy of Science,* ed. L. H. Nelson and J. Nelson, pp. 121–138. Dordrecht: Kluwer Academic Publishers.

Potts, W. K. 2002. Wisdom through immunogenetics. *Nature Genetics* 30 (February): 130–131.

Potts, W. K., C. J. Manning, and E. K. Wakeland. 1991. Mating patterns in seminatural populations of mice influenced by MHC genotype. *Nature* 352:619–621.

Pugh, G. 1977. *Biological Origins of Human Values.* New York: Basic Books.

Qidwai, W. 2000. Perceptions about female sexuality among young Pakistani men presenting to family physicians at a teaching hospital in Karachi. *Journal of the Pakistan Medical Association* 50 (2): 74–77.

Raboch, J., and V. Barták. 1983. Coitarche and orgastic capacity. *Archives of Sexual Behavior* 12 (5): 409–413.

Raboch, J., and J. Raboch. 1992. Infrequent orgasms in women. *Journal of Sex and Marital Therapy* 18 (2): 114–120.

Rancour-Laferriere, D. 1983. Four adaptive aspects of the female orgasm. *Journal of Social and Biological Structures* 6:319–333.

Reeve, H. K., and P. W. Sherman. 1993. Adaptation and the goals of evolutionary research. *Quarterly Review of Biology* 68 (1): 1–32.

Reichenbach, H. 1966. *The Rise of Scientific Philosophy.* Berkeley: University of California Press.

Reusch, T., M. A. Häberli, P. B. Aeschlimann, and M. Milinski. 2001. Female sticklebacks count alleles: a new strategy of sexual selection explaining MHC polymorphism. *Nature* 414:300–302.

Reznick, D., and J. Travis. 1996. The empirical study of adaptation in natural populations. In *Adaptation,* ed. M. R. Rose and G. V. Lauder, pp. 243–291. San Diego: Academic Press.

Richardson, R. 1984. Biology and ideology: the interpretation of science and values. *Philosophy of Science* 51:396–420.

Ridley, M. 1996. *Evolution*. Cambridge, Eng.: Blackwell Science.

Robbins, M., and G. Jensen. 1977. Multiple orgasm in males. In *Progress in Sexology*, ed. R. Gemme and C. C. Wheeler, pp. 323–328. New York: Plenum Press.

Robinson, P. A. 1976. *The Modernization of Sex*. New York: Harper & Row.

Rodgers, J. E. 2001. *Sex: A Natural History*. New York: Times Books–Henry Holt.

Rodriguez, V. 1995. Function of the spermathecal muscle in *Chelymorpha alternans Boheman*. *Physiological Entomology* 19:198–202.

Rooney, P. 1992. On values in science: is the epistemic/non-epistemic distinction useful? In *Proceedings of the Biennial Meeting of the Philosophy of Science Association, 1992*, vol. 1, pp. 13–22. East Lansing, MI: Philosophy of Science Association.

Rose, K. D. 2001. Edentata and Pholidota (armadillos, anteaters, and tree sloths). In *Nature Encyclopedia of Life Sciences*. London: Nature Publishing Group; http://www.els.net/[doi:10.1038/npg.els.0001556].

Rose, M. R., and G. V. Lauder. 1996. *Adaptation*. San Diego: Academic Press.

Rosenthal, H. C. 1951. Sex habits of European women vs. American women. *Pageant* 7:52–59.

Rosvald, H. E. 1954. Sexual behavior in subhuman primates. Paper presented at the symposium Genetic, Psychological, and Hormonal Factors in the Establishment and Maintenance of Patterns of Sexual Behavior in Mammals, Amherst MA, November 17–19, 1954.

Rowell, T. E. 1963. Behaviour and reproductive cycles of female macaques. *Journal of Reproduction and Fertility* 6:193–203.

———1972. Female reproductive cycles and social behaviour in primates. *Advances in the Study of Behavior* 4:69–105.

Rucknagel, D. L., and J. V. Neel. 1961. The hemoglobinopathies. In *Progress in Medical Genetics*, ed. A. G. Steinberg, vol. 1, pp. 158–260. New York: Grune and Stratton.

Saayman, G. 1970. The menstrual cycle and sexual behavior in a troop of free-ranging chacma baboons *(Papio ursinus)*. *Folia Primatologica* 12:81–110.

———1975. The influence of hormonal and ecological factors upon sexual behavior and social organization in old world primates. In *Socioecology and Psychology of Primates,* ed. R. Tuttle, pp. 181–204. The Hague: Mouton.

Schiavi, R. C., and R. T. Segraves. 1995. The biology of sexual function. *The Psychiatric Clinics of North America* 18:7–23.

Schluter, D. 1988. Estimating the form of natural selection on a quantitative trait. *Evolution* 42:849–861.

Schmitt, F. F., ed. 1994. *Socializing Epistemology: The Social Dimensions of Knowledge.* Studies in Epistemology and Cognitive Theory. Lanham, MD: Rowman & Littlefield.

Schnabl, S. 1980. Correlations and determinants of functional sexual disturbances. In *Medical Sexology,* ed. R. Forleo and W. Pasini, pp. 153–161. Littleton, MA: PSG Publishing.

Seger, J., and J. W. Stubblefield. 1996. Optimization and adaptation. In *Adaptation,* ed. M. R. Rose and G. V. Lauder, pp. 93–125. San Diego: Academic Press.

Serjeant, G. R. 2001. The emerging understanding of sickle cell disease. *British Journal of Haematology* 112:3–18.

Shaffer, N. E. 1996. Understanding bias in scientific practice. *Philosophy of Science* 63:S89-S97.

Sherfey, M. J. 1973. *The Nature and Evolution of Female Sexuality.* New York: Random House.

Sherman, P. 1988. The levels of analysis. *Animal Behaviour* 36 (2): 616–619.

———1989. The clitoris debate and the levels of analysis. *Animal Behaviour* 37 (4): 697–698.

Shope, D. F. 1968. Orgastic responsiveness of selected college females. *Journal of Sex Research* 4 (3): 206–219.

Short, R. V. 1997. Review of R. R. Baker and M. A. Bellis, *Human Sperm Competition: Copulation, Masturbation, and Infidelity. European Sociobiological Society* 47: 20–23.

Simmons, L. W. 2001. *Sperm Competition and Its Evolutionary Consequences in the Insects.* Princeton: Princeton University Press.

Sinervo, B., and A. L. Basolo. 1996. Testing adaptation using phenotypic manipulations. In *Adaptation,* ed. M. R. Rose and G. V. Lauder, pp. 149–187. San Diego: Academic Press.

Singer, I. 1973. *The Goals of Human Sexuality.* New York: W. W. Norton.

Singer, J., and I. Singer. 1972. Types of female orgasm. *Journal of Sex Research* 8 (4): 255–267.

Singh, D., W. Meyer, R. J. Zambarano, and D. F. Hulbert. 1998. Frequency and timing of coital orgasm in women desirous of becoming pregnant. *Archives of Sexual Behavior* 27 (1): 15–29.

Singh, S. R., B. N. Singh, and H. F. Hoenigsberg. 2002. Female remating, sperm competition, and sexual selection in *Drosophila. Genetic Molecular Research* 1 (3): 178–215.

Slater, E., and M. Woodside. 1951. *Patterns of Marriage. A Study of Marriage Relationships in the Urban Working Classes.* London: Cassell.

Slob, A. K., W. Groeneveld, and J. J. van der Werff ten Bosch. 1986. Physiological changes during copulation in male and female stumptail macaques *(Macaca arctoides). Physiology and Behavior* 38:891–895.

Slob, A. K., and J. J. van der Werff ten Bosch. 1991. Orgasm in non-human species. In *Proceedings of the First International Conference on Orgasm,* pp. 135–149. Bombay, India: VRP Publishers.

Smith, J. M. 1978. The concepts of sociobiology. In *Morality as a Biological Phenomenon: Report of the Dahlem Workshop on Biology and Morals, Berlin 1977,* ed. G. S. Stent, pp. 23–25. Berlin: Abakon Verlagsgesellschaft.

Smith, R. L. 1984. Human sperm competition. In *Sperm Competition and the Evolution of Animal Mating Systems,* ed. R. L. Smith, pp. 601–659. London: Academic Press.

Sober, E. 1993. *Philosophy of Biology.* San Francisco: Westview Press.

Solomon, M. 1995. Multivariate models of scientific change. *Proceedings of the Biennial Meeting of the Philosophy of Science Association, 1994,* ed. R. Burian, M. Forbes, and D. Hull, pp. 287–297. East Lansing, MI: Philosophy of Science Association.

———2001. *Social Empiricism.* Cambridge: MIT Press.

Spector, I. P., and M. P. Carey. 1990. Incidence and prevalence of the sexual dysfunctions: a critical review of the empirical literature. *Archives of Sexual Behavior* 19 (4): 389–408.

Stanley, L. 1995. *Sex Surveyed, 1949–1994: From Mass Observation's 'Little Kinsey' to the National Survey and the Hite Reports.* London: Taylor & Francis.

Stoddart, D. M. 1991. *The Scented Ape: The Biology and Culture of Human Odour.* Cambridge: Cambridge University Press.

Stone, H. M., and A. S. Stone. 1952. *A Marriage Manual: A Practical Guidebook to Sex and Marriage.* New York: Simon and Schuster.

Strassman, B. 1981. Sexual selection, parental care, and concealed ovulation in humans. *Ethology and Sociobiology* 2:31–40.

Streitfeld, D. 1988. Shere Hite and the trouble with numbers. *Chance* 1 (3): 26–31.

Suggs, R. C. 1971. Sex and personality in the Marquesas: a discussion of the Linton-Kardiner report. In *Human Sexual Behavior: Variations in the Ethnographic Spectrum,* ed. D. S. Marshall and R. C. Suggs, pp. 163–186. Englewood Cliffs, NJ: Prentice-Hall.

Suggs, R. C., and D. S. Marshall. 1971. Anthropological perspectives on human sexual behavior. In *Human Sexual Behavior: Variations in the Ethnographic Spectrum,* ed. D. S. Marshall and R. C. Suggs, pp. 218–243. Englewood Cliffs, NJ: Prentice-Hall.

Swan, S. H. 2003. Do environmental agents affect semen quality? *Epidemiology* 14:261–262.

Symons, D. 1979. *The Evolution of Human Sexuality.* New York: Oxford University Press.

———1980a. Precis of the evolution of human sexuality. *Behavioral and Brain Sciences* 3:171–181.

———1980b. Response to commentaries. *Behavioral and Brain Sciences* 3:203–211.

———1990. Adaptiveness and adaptation. *Ethology and Sociobiology* 11:427–444.

Tavris, C., and S. Sadd. 1977. *The Redbook Report on Female Sexuality.* New York: Delacorte Press.

Taylor, E. S. 1962. *Essentials of Gynecology.* Philadelphia: Lea & Febiger.

Temerlin, M. K. 1975. *Lucy: Growing Up Human.* Palo Alto: Science and Behavior Books.

Templeton, A. R. 1982. Adaptation and the integration of evolutionary forces. In *Perspectives on Evolution,* ed. R. Milkman, pp. 15–31. Sunderland, MA: Sinauer.

Terman, L. M. 1938. *Psychological Factors in Marital Happiness.* New York: McGraw-Hill.

———1951. Correlates of orgasm adequacy in a group of 556 wives. *Journal of Psychology* 32: 115–172.

Terry, J. 1999. *An American Obsession: Science, Medicine, and the Place of Homosexuality in Modern Society.* Chicago: University of Chicago Press.

Thomas, M. L., J. H. Harger, D. K. Wagener, B. S. Rabin, and T. J. Gill III. 1985. HLA sharing and spontaneous abortion in humans. *American Journal of Obstetrics and Gynecology* 151:1053–1058.

Thornhill, R. 1990. The study of adaptation. In *Interpretation and Explanation in the Study of Animal Behavior,* vol. 2: *Explanation, Evolution, and Adaptation,* ed. M. Bekoff and D. Jamieson, pp. 31–62. Boulder: Westview Press.

Thornhill, R., S. W. Gangestad, and R. Comer. 1995. Human female orgasm and mate fluctuating asymmetry. *Animal Behaviour* 50: 1601–1615.

Tinbergen, N. 1963. On aims and methods of ethology. *Zeitschrift für Tierpsychologie* 20:410–433.

Tooby, J., and I. DeVore. 1987. The reconstruction of hominid behavioral evolution through strategic modeling. In *The Evolution of Human Behavior: Primate Models,* ed. W. G. Kinzey, pp. 183–237. Albany, NY: SUNY Press.

Vance, E. B., and N. N. Wagner. 1976. Written descriptions of orgasms: a study of sex differences. *Archives of Sexual Behavior* 5:87–89.

Wade, M. J. 1987. Measuring sexual selection. In *Sexual Selection: Testing the Alternatives,* ed. J. W. Bradbury and M. B. Andersson, pp. 197–207. New York: John Wiley.

Wade, M. J., and S. Kalisz. 1990. The causes of natural selection. *Evolution* 44:1947–1955.

Wallin, P. 1960. A study of orgasm as a condition of women's enjoyment of intercourse. *Journal of Social Psychology* 51: 191–198.

Wasser, S. K. 1983. Reproductive competition and cooperation among female yellow baboons. In *The Social Behavior of Female Vertebrates,* ed. S. K. Wasser, pp. 349–390. New York: Academic Press.

Wasser, S. K., and M. L. Waterhouse. 1983. The establishment and maintenance of sex biases. In *The Social Behavior of Female Vertebrates,* ed. S. K. Wasser, pp. 19–35. New York: Academic Press.

Wedekind, C., and S. Füri. 1997. Body odour preferences in men and women: do they aim for specific MHC combinations or simply heterozygosity? *Proceedings of the Royal Society of London* B 264:1471–1479.

Wedekind, C., T. Seebeck, F. Bettens, and A. J. Paepke. 1995. MHC-dependent mate preferences in humans. *Proceedings of the Royal Society of London* B 260:245–249.

West-Eberhard, M. J. 1987. Conflicts between and within the sexes in sexual selection: group report. In *Sexual Selection: Testing the Alternatives,* ed. J. W. Bradbury and M. B. Andersson, pp. 181–195. New York: John Wiley.

———1992. Adaptation. In *Keywords in Evolutionary Biology,* ed. E. F. Keller and E. A. Lloyd, pp. 13–18. Cambridge: Harvard University Press.

———2003. *Developmental Plasticity and Evolution.* Oxford: Oxford University Press.

Whipple, B., W. E. Hartman, and M. A. Fithian. 1994. Orgasm. In *Human Sexuality: An Encylopedia,* ed. V. L. Bullough and B. Bullough, pp. 430–433. New York: Garland Publishing.

Wildt, L., S. Kissler, P. Licht, and W. Becker. 1998. Sperm transport in the human female genital tract and its modulation by oxytocin as assessed by hysterosalpingoscintigraphy, hysterotonography, electrohysterography and Doppler sonography. *Human Reproduction Update* 4 (5): 655–666.

Williams, G. C. 1985. A defense of reductionism in evolutionary biology. *Oxford Surveys in Evolutionary Biology* 2:1–27.

Wilson, E. O. 1975. *Sociobiology: The New Synthesis.* Cambridge: Harvard University Press.

———1978. *On Human Nature.* Cambridge: Harvard University Press.

Wimsatt, W. 1987. False models as a means to truer theories. In *Neutral Models in Biology,* ed. M. H. Nitecki and A. Hoffman, pp. 23–55. Oxford: Oxford University Press.

Wolfe, L. 1979. Behavioral patterns of estrous females of the Arashiyama West troop of Japanese macaques *(Macaca fuscata). Primates* 20 (4): 525–534.

———1984. Japanese macaque female sexual behavior: a comparison of Arashiyama East and West. In *Female Primates: Studies by Women Primatologists,* ed. M. F. Small, pp. 141–157. New York: Alan R. Liss.

———1991. Human evolution and the sexual behavior of female primates. In *Understanding Behavior: What Primate Studies Tell Us about Human Behavior,* ed. J. D. Loy and C. B. Peters, pp. 121–151. New York: Oxford University Press.

Woodside, M. 1950. *Sterilization in North Carolina. A Sociological and Psychological Study.* Chapel Hill: University of North Carolina Press.

Wrangham, R. W. 1980. An ecological model of female-bonded primate groups. *Behaviour* 75:262–300.

Wrangham, R. W., and D. Peterson. 1996. *Demonic Males: Apes and the Origins of Human Violence.* Boston: Houghton Mifflin.

Yarros, R. S. 1933. *Modern Woman and Sex: A Feminist Physician Speaks.* New York: Vanguard Press.

Yerkes, R. 1939a. Social dominance and sexual status in the chimpanzee. *Quarterly Review of Biology* 14 (2): 115–136.

———1939b. Sexual behavior in the chimpanzee. *Human Biology* 11:78–110.

Zaviacic, M., J. Jakubovsky, S. Polák, A. Zaviacicová, I. K. Holomán, J. Blazekova, and P. Gregor. 1984. The fluid of female urethral expulsions analysed by histochemical electron-microscopic and other methods. *Histochemical Journal* 16:445–447.

Zaviacic, M., and B. Whipple. 1993. Update on the female prostate and the phenomenon of female ejaculation. *Journal of Sex Research* 30 (2): 148–151.

Zaviacic, M., A. Zaviacicová, I. K. Holomán, and J. Molcan. 1988. Female urethral expulsions evoked by local digital stimulation of the G-spot: differences in response patterns. *Journal of Sex Research* 24:311–318.

Zuckerman, S. 1932. *The Social Life of Monkeys and Apes.* New York: Harcourt, Brace.

Zuk, M. 2002. *Sexual Selections: What We Can and Can't Learn about Sex from Animals.* Berkeley: University of California Press.

Zumpe, D., and R. P. Michael. 1968. The clutching reaction and orgasm in the female rhesus monkey *(Macaca mulatta). Journal of Endocrinology* 40:117–123.

# Acknowledgments

This book would not have been written without the help, criticism, and support of many people.

I am always grateful to Bas van Fraassen and Dick Jeffrey, who served as my mentors and supporters during this long project. I am especially saddened by the fact that Dick will not see the book in print. I also have a special debt to Dick Lewontin, who hosted me when I was a Research Associate at the Museum of Comparative Zoology at Harvard during the fall of 1989, the spring of 1998, and the spring of 2001. His critical comments were invaluable in formulating my views.

I am indebted to Libby Prior for asking the question that got this research project started.

Many people have helped shape my thoughts regarding the intersection of evolution, philosophy of science, feminism, and female sexuality. These include: Jerome Barkow, Jim Bogen, Gillian Brown, David Buller, Dick Burian, Nancy Cartwright, Helena Cronin, Rachel Davis, Richard Dawkins, Persi Diaconis, Ford Doolittle, John Dupré, Marc Ereshefsky, Anne Fausto-Sterling, Linda Fedigan, Marc Feldman, Arthur Fine, Marilyn Fithian, Richard Francis, Herb Gintis, Peter Godfrey-Smith, Deborah Gordon, Monica Green, Jim Griesemer, Paul Griffiths, Susan Hale, Donna Haraway, William Hartman, Sally Haslanger, Chris Horvath, Sarah

Blaffer Hrdy, Martin Jones, Eva and Jeffrey Kittay, Louise Knapp, Sally Kornblith, Kevin Laland, Elaine Landry, Tom Laqueur, Bruno Latour, Helen Longino, Alan Love, Laura Lovett, Judith Masters, Alison McIntyre, Christia Mercer, Sandra Mitchell, Lynn Nelson, Maggie Ostler, Anne Pyburn, Ken Reisman, Bob Richards, Jonathan Sills, Louise Silvern, Elliott Sober, Hamish Spencer, Don Symons, Bob Trivers, Kent Van Cleave, Bill Wimsatt, Rasmus Winther, Juliet Wittman, Alison Wylie, and Marlene Zuk. I am sure there are some I have forgotten, and I ask their forgiveness.

I am especially grateful to Steve Downes, Melinda Fagan, David Hull, Michael Dietrich, Stephen Grover, Thalia Brine Schlossberg, and Richmond Campbell, who read entire drafts of the manuscript. Their extensive comments and suggestions for revisions are deeply appreciated.

I would like to give special thanks to three statisticians on whose expertise I relied, Persi Diaconis, Marcus Feldman, and Stuart Lloyd. They variously provided independent evaluations of the statistical evidence in Chapter 7 prior to my analysis, and confirmed and reinforced my own detailed statistical analysis of that evidence.

I owe a particular debt to my research assistants who helped me with this book: Carl Anderson, Mark Borello, Joanne Chen, Sarah Hahn, Anne Mylott, Greg Ray, and Eric Schwitzgebel. I would like to single out Stephen Crowley and Alex Klein for their cheerful and attentive presence, and their help with the push to finish the book.

Having worked in three departments during the writing of this book, I have many people to thank. I am sincerely grateful to my colleagues at the University of California, San Diego, University of California, Berkeley, and Indiana University for their support. There are a number of people who deserve to be recognized for their help and comments, including Steve Stich, Patty and Paul Churchland, and Philip Kitcher from UCSD, and Hans Sluga, Bert Dreyfus, and Janet Broughton from Berkeley.

There are many colleagues from Indiana University from whom I have learned a great deal. This book has profited from the insights of my colleagues in and out of the History and Philosophy of Science Department, including especially Jordi Cat, Lynda Delph, Mark Kaplan, Ellen Ketterson, Noretta Koertge, Curt Lively, Mike Wade, and Joan Weiner. The recent Director of the Kinsey Institute, John Bancroft, was a special help to me.

This material served as a basis for many presentations to scholarly audiences over the years. I owe a great debt to Evelyn Fox Keller for encouraging me to pursue, in the very early stages, this case study, and for bringing it to the attention of Elizabeth Potter, who organized a conference at Hamilton College, at which I first presented my research on the topic to critical colleagues in September 1985. I owe Libby Potter my deepest thanks for this early forum. In addition, I am especially grateful to the American Philosophical Association, the University of Auckland (New Zealand), the University of Calgary, the University of Chicago, the Ecole Nationale Supérieure des Mines (Paris), the Ecole Polytechnique–CNRS (Paris), the Kinsey Institute, Northwestern, Oxford, Princeton, Rutgers, Stanford, and MIT, where I presented my work. I am also grateful for the coverage that Karen Wright gave my views in *Discover* magazine. I was privileged to be a part of the Workshop on Philosophy of Biology at the University of Pittsburgh, and especially appreciate the commentary given by Karen Arnold. Gordon McOuat generously arranged for me to meet with the Evolution Study Group at King's University College and Dalhousie University, from whom I received extremely valuable comments. I am further indebted to Gordon for his critical advice and observations during the writing of the book.

Gordon Getty and the Gruter Institute for Biology and the Law provided me with opportunities to discuss my work with a variety of people, including Frans de Waal. I am especially indebted to

Frans, Shirley Strum, and Richard Wrangham for insights about the apes and monkeys that they study; I cannot imagine having better guides.

Sally Haslanger's, Evelyn Fox Keller's, and Helen Longino's courage in pursuing feminist thought in philosophy has been an inspiration to me. I am also very grateful to have my first published foray into the topic of female orgasm (1993) reprinted in the text *Feminism and Science*, edited by Evelyn Fox Keller and Helen Longino (1996). This essay, which was the text of a lecture at the Pacific American Philosophical Association, was published because it was chosen as among the ten best talks of that conference.

I wish to thank the National Science Foundation for a Scholar's Grant to work on this project and the University of California Committee on Research for financial support from 1986 through 1997. I have a special debt to Dean Kumble Subbaswamy for making my life easier. My deepest gratitude goes to Arnold and Maxine Tanis, who created my endowed chair; their material and moral support has been essential to the project.

I have been especially lucky to have Lindsay Waters as my editor. His patience, guidance, and enthusiasm inspired and sustained me through illness and trial.

This book would not have been written without the attention and support of Steve Gould. His adoption of my work proved to be the push that I needed in order to think of the project as a book. I am eternally grateful for his respect, and deeply regret that he will not see the final version of the story. Finally, without the love, kindness, and moral support of my parents, Dr. Stuart and Ruth Lloyd, the book would not exist.

# Index

account of female orgasm (Morris, Gallup, and Suarez)
Mould, Douglas, 21

Newton, Niles, 50, 62, 104
Nipples, male. *See* Byproduct account of male nipples
Nonadaptive accounts of female orgasm. *See* Byproduct account of female orgasm; Hamburg, Beatrix
Nonhuman primate female orgasm: chimpanzees, 55, 64, 128; rhesus macaques, 55, 99–100, 120, 126–127; bonobos, 64, 128, 234; Japanese macaques, 87, 117, 120, 127; Hrdy on, 100–101; anatomical differences with humans, 137–139, 263n5. *See also* Clutch reaction; Stumptail macaques
Nonhuman primate sexual behavior and hormonal status, 7, 55, 118
Nonpair-bond adaptive accounts of female orgasm. *See* Female signaling to male (Alexander and Noonan); Assisting male ejaculation (Allen and Lemmon); Intermittent reinforcement (Diamond; Hrdy); Orgasm induces abortion (Bernds and Barash); Promiscuity protection (Hrdy); Reproductive health (Sherfey). *See also* Pair-bond adaptive accounts of female orgasm
Noonan, Katherine. *See* Female signaling to male account of female orgasm (Alexander and Noonan)

Ober, Carole, 9
Objectivity: evaluation of science, 220, 239; relation of data to theory, 241; objective inquiry, 245–255, 256

Orgasm induces abortion account of female orgasm (Bernds and Barash), 83–86
Orgasm and intercourse distinguished, 2, 14–15, 28–34, 61, 85–89, 223–225; agreement on distinction, 25; explanation of distinction, 37–38, 67; distinction and pair-bond accounts of female orgasm, 44–106 passim; Hamburg on, 70–71; Rancour-Laferriere on, 72, 74–75; Alcock on, 74–75; distinction as recent phenomenon, 77–83, 115, 224–225; Allen and Lemmon on, 79–83; Symons and Gould on, 110–112; in cross-cultural reports, 115–116. *See also* Clitoris as an adaptation; Masturbation, female orgasm by; Orgasm and intercourse not distinguished
Orgasm and intercourse not distinguished, 224; by Morris, 51–53, 57, 59; by Pugh, 61; by Crook, 61–62; by Newton, 62; by Eibl-Eibesfeldt, 62–63; by Campbell, 66–67; by Beach, 68, 70; by Allen and Lemmon, 79–83, 85–89; adaptation and androcentrism, 239. *See also* Orgasm and intercourse distinguished
Orgasm versus sexual excitement: physiologically distinct, 38–39; and reproductive success, 66–67; Hrdy on, 99–101; Hite on, 101; Jamieson on, 162, 277–278. *See also* Clitoris as an adaptation; Promiscuity protection account of female orgasm (Hrdy)
Oxytocin, 23, 157, 185–190

Pair bond: definition, 49; and female orgasm, 49–76

159, 163, 165, 172; Jamieson on, 162; Alcock and Sherman on, 163; not linked through uterine suction, 181–182, 185, 229, 259; Baker and Bellis on, 208. *See also* Reproductive success and female orgasm linked

Reznick, David, 7

Rodgers, J. E., 16, 216

Rodriquez, V., 194

Rosenthal, H. C., 32

Sadd, S., 33

Samples, unrepresentative: insufficient for evolutionary purposes, 36, 75, 78–79, 89–91, 93–94, 98–99, 113–114, 131–136, 221; Masters and Johnson, 52–53, 114–115, 225; Alcock and Rancour-Laferriere, 75; Bartlett and Allen and Lemmon, 78–79; Diamond and Hrdy, 89–91, 93–95; Hrdy, 98–99, 146; Fox and Fox, 183, 193; Thornhill et al., 212. *See also* Methodological problems of sexology research

Schnabl, S., 32, 35, 98

Science, 18, 20; Longino's criteria, 246–255; importance of impartiality, 244–248, 256

Secondary adaptation, female orgasm as: Hrdy on, 97; lack of research on, 146–147; Gould on, 164

Selective samples. *See* Samples, unrepresentative

Selective use of evidence. *See* Evidence, selective use of

Sex differences in sexuality. *See* Male and female sexuality compared

Sexual excitement versus orgasm. *See* Orgasm versus sexual excitement

Sexy males make good fathers claim: Alcock, 73, 147, 151; Rancour-Laferriere, 73–74, 100; unsupported, 74, 225

Sherfey, Mary Jane, 92–95, 104, 114, 225. *See also* Reproductive health account of female orgasm (Sherfey)

Sherman, Paul, 150, 156, 158–166, 169–177, 252

Shope, David, 52

Short, Roger, 194–195, 247, 249

Sinervo, Barry, 4, 12

Singer, Irving, 183–185, 228

Singer, J., 183–184

Singh, D., 16, 216

Slater, E., 33

Slob, A., 118–119, 123–125, 129, 227–228

Smith, Robert, 179–180, 195–198, 211

Social consequences of views of female orgasm, 18–19, 44, 114, 142–143

Sociobiology, 48–49

Solomon, Miriam, 220, 244

Sperm competition: appeal to adaptationists, 4–5, 7, 131, 134, 217, 219; Smith on sperm manipulation, 16, 173, 179, 197, 216–219, 298–299; popularity of account, 16–17, 175, 216–217, 249–250; in insects, 193–194; "good genes" and "sexy sons" hypotheses, 194–195, 198; problems with, 194–195, 197–198, 218; advantages to females, 195. *See also* Uterine-upsuck account of female orgasm (Baker and Bellis)

Stanley, L., 27, 33

Stone, A. S., 33

Stone, H. M., 33

Strassman, B., 102